Big Data Analytics in Future Power Systems

Big Data Analytics in Future Power Systems

Edited by
Ahmed F. Zobaa and Trevor J. Bihl

CRC Press
Taylor & Francis Group
Boca Raton London New York

CRC Press is an imprint of the
Taylor & Francis Group, an **informa** business

CRC Press
Taylor & Francis Group
6000 Broken Sound Parkway NW, Suite 300
Boca Raton, FL 33487-2742

First issued in paperback 2020

© 2019 by Taylor & Francis Group, LLC
CRC Press is an imprint of Taylor & Francis Group, an Informa business

No claim to original U.S. Government works

ISBN 13: 978-0-367-73338-4 (pbk)
ISBN 13: 978-1-138-09588-5 (hbk)

Library of Congress Cataloging-in-Publication Data

Names: Zobaa, Ahmed F., editor. | Bihl, Trevor J., editor.
Title: Big data analytics in future power systems / [edited by] Ahmed F. Zobaa and Trevor J. Bihl.
Description: Boca Raton : Taylor & Francis, a CRC title, part of the Taylor & Francis imprint, a member of the Taylor & Francis Group, the academic division of T&F Informa, plc, 2018. | Includes bibliographical references.
Identifiers: LCCN 2018024681 | ISBN 9781138095885 (hardback : acid-free paper) | ISBN 9781315105499 (ebook)
Subjects: LCSH: Smart power grids—Data processing. | Big data. | Electric power systems.
Classification: LCC TK3105 .B54 2018 | DDC 621.310285/57--dc23
LC record available at https://lccn.loc.gov/2018024681

Visit the Taylor & Francis Web site at
http://www.taylorandfrancis.com

and the CRC Press Web site at
http://www.crcpress.com

Contents

Preface vii

Acknowledgments ix

Editors xi

List of Contributors xiii

1. Introduction 1
   Ahmed F. Zobaa and Trevor J. Bihl

2. Big Data Application and Analytics in a Large-Scale Power System....9
   Jeremy Lin, Elham Foruzan, and Fernando H. Magnago

3. The Role of Big Data in Smart Grid Communications 37
   Francisco M. Portelinha Júnior and Denisson Q. Oliveira

4. Big Data Optimization in Electric Power Systems: A Review 55
   Iman Rahimi, Abdollah Ahmadi, Ahmed F. Zobaa, Ali Emrouznejad, and Shady H.E. Abdel Aleem

5. Security Methods for Critical Infrastructure Communications 85
   Ahmed F. Zobaa and Trevor J. Bihl

6 Data-Mining Methods for Electricity Theft Detection 107
   Trevor J. Bihl and Ahmed F. Zobaa

7. Unit Commitment Control of Smart Grids 125
   Salam Hajjar

8. A New Transformer Differential Protection Algorithm Based on Data Pattern Recognition 143
   Ernesto Vázquez Martínez, Héctor Esponda Hernández, and Manuel A. Andrade Soto

Index 169

Preface

The increasing penetration of the Smart Grid, the desire to monitor all components in the power grid, and the expansion of the Internet of Things (IoT) have resulted in Big Data problems throughout power systems. In general, Big Data is permeating all aspects of our lives today and is a result of the improvements of sensors and their availability, expanding communication abilities standards, and ever increasing abilities to store digital data. Inherently, Big Data is created when data are logged or collected at very high rates (velocity) on any number of processes (variety) with as fine of detail possible (volume). The result of the ability to collect endless data is the emergence of Big Data. However, power systems are connected to physical devices and critical infrastructure (CI) and thus additional research problems and concerns exist in power system Big Data.

Big Data Analytics in Future Power Systems aims to discuss Big Data problems and solutions inherent in future power systems. It thus introduces methods available to handle and make sense of Big Data in power systems. This book covers a wide range of power system topics, from metering to transformer monitoring. Demand prediction and planning under uncertain generation, as seen with renewables, are further shown to be enabled by the wealth of data available in Big Data. Additionally, this book discusses the various security concerns that become manifest with Big Data and expanded communications in power grids and CI.

It introduces the concepts, methods, and approaches needed by power system professionals to improve their understanding of Big Data challenges and capabilities. Further, it provides a glimpse of future directions of Big Data in power systems. The book is composed of a collection of carefully selected and reviewed chapters written by diverse experts in the field.

Ahmed F. Zobaa
Brunel University London
United Kingdom

Trevor J. Bihl
Wright State University
United States

Acknowledgments

In addition to the authors themselves, we would like to thank the following external researchers, professionals, and faculty who provided their time and effort in reviewing chapters of this book. Additional thanks go to Kyra Lindholm and Vanessa Garrett at Taylor & Francis Group/CRC Press for courtesy, professionalism, and support in this endeavor.

Tim Carbino
Air Force Institute of Technology, USA

James Cordiero
University of Dayton, USA

Parisa Fatheddin
Air Force Institute of Technology, USA

Mark A Friend
Northern Arizona University, USA

Jordan Goldmeier
Cambia Factor, USA

Salam Hajjar
Marshall University, USA

Teresa Hawkes
University of Oklahoma, USA

Ronnie Minhaz
TC Services Inc

Todd Paciencia
Independent Researcher, USA

Carl Parson
Scientific Test and Analysis Techniques Center of Excellence, USA

Francisco Martins Portelinha Junior
National Institute of Telecommunications, Brazil

Daniel Steeneck
Air Force Institute of Technology, USA

David Smalenberger
Independent Researcher, USA

Editors

Ahmed F. Zobaa received his BSc (Hons), MSc, and PhD degrees in electrical power and machines from Cairo University, Egypt, in 1992, 1997, and 2002, respectively. He received his postgraduate certificate in Academic Practice from the University of Exeter, UK in 2010. In addition, he received the Doctoral of Science from Brunel University London, UK in 2017. He was an instructor during 1992–1997, a teaching assistant during 1997–2002, and an assistant professor during 2002–2007 at Cairo University, Egypt. From 2007 to 2010, he was a senior lecturer in renewable energy at University of Exeter, UK. Currently, he is a senior lecturer in electrical and power engineering, an MSc Course Director, and a Full Member of the Institute of Energy Futures at Brunel University London, UK. His main areas of expertise include power quality, (marine) renewable energy, smart grids, energy efficiency, and lighting applications.

Ahmed F. Zobaa is an executive editor for the *International Journal of Renewable Energy Technology*. He is an editor-in-chief for *Technology and Economics of Smart Grids and Sustainable Energy*, and *International Journal of Electrical Engineering Education*. He is also an editorial board member, editor, associate editor, and editorial advisory board member for many international journals. He is a registered chartered engineer, chartered energy engineer, European engineer, and international professional engineer. He is also a registered member of the Engineering Council UK, Egypt Syndicate of Engineers, and the Egyptian Society of Engineers. He is a senior fellow of the Higher Education Academy of UK. He is a fellow of the Institution of Engineering and Technology, the Energy Institute of UK, the Chartered Institution of Building Services Engineers, the Institution of Mechanical Engineers, the Royal Society of Arts, the African Academy of Science, and the Chartered Institute of Educational Assessors. He is a senior member of the Institute of Electrical and Electronics Engineers. In addition, he is a member of the International Solar Energy Society, the European Power Electronics and Drives Association, and the IEEE Standards Association.

Trevor J. Bihl received a PhD degree in electrical engineering from the Air Force Institute of Technology, Wright Patterson AFB, OH. Additionally, he received the BS and MS degrees in electrical engineering from Ohio University, Athens, OH. Primarily, he is a research scientist and engineer. He is also an educator and holds faculty positions at Wright State University in both the Department of Biomedical, Industrial and Human Factors Engineering and the Department of Pharmacology & Toxicology. His main areas of expertise include statistical data analysis, pattern recognition, communication systems, autonomous systems, cyber security, operations research, and remote sensing.

Trevor J. Bihl is an associate editor for the *International Journal of Electrical Engineering Education*. He is also the author of *Biostatistics Using JMP: A Practical Guide*. He is a member of the Institute of Electrical and Electronics Engineers (IEEE), and the Institute for Operations Research and the Management Sciences (INFORMS). Also, he is a member of the INFORMS Subdivision Council.

List of Contributors

S. H. E. Abdel Aleem
Mathematical, Physical and
  Engineering Sciences
  Department
15th of May Higher Institute of
  Engineering
Cario, Egypt

A. Ahmadi
School of Electrical Engineering and
  Telecommunications
University of New South Wales
Kensington, New South Wales,
  Australia

M. Andrade
Mechanical and Electrical
  Engineering Faculty
Universidad Autónoma de Nuevo
  León
San Nicolás de los Garza, Mexico

T. J. Bihl
Department of Biomedical,
  Industrial & Human Factors
  Engineering
Department of Pharmacology &
  Toxicology
Wright State University
Dayton, Ohio

A. Emrouznejad
Aston Business School
Aston University
Birmingham, United Kingdom

H. Esponda
Mechanical and Electrical
  Engineering Faculty
Universidad Autónoma de Nuevo
  León
San Nicolás de los Garza, Mexico

E. Foruzan
Department of Electrical &
  Computer Engineering
University of Nebraska-Lincoln
Lincoln, Nebraska

S. Hajjar
Weisberg Division of Engineering
Marshall University
Huntington, West Virginia

J. Lin
Transmission Analytics
Austin, Texas

F. H. Magnago
Faculty of Engineering
Universidad de Rio Cuarto
Río Cuarto, Argentina

and

Nexant, Inc.
Río Cuarto, Argentina

D. Q. Oliveira
Computer Engineering
  Department
Federal University of Maranhão
São Luís, Brazil

F. M. Portelinha Júnior
Electrical Engineering
  Department National Institute
  of Telecommunications
  (INATEL)
Santa Rita do Sapucaí, Brazil

I. Rahimi
Young Researchers and Elite Club
  Isfahan (Khorasgan) Branch
Islamic Azad University
Isfahan, Iran

E. Vázquez
Mechanical and Electrical
  Engineering Faculty
Universidad Autónoma de Nuevo
  León
San Nicolás de los Garza, Mexico

A. F. Zobaa
Electronic and Computer
  Engineering Department
Brunel University London
London, United Kingdom

1

Introduction

Ahmed F. Zobaa
Brunel University London

Trevor J. Bihl
Wright State University

CONTENTS

1.1   Introduction...1
1.2   Big Data 2
1.3   Future Power Systems 2
1.4   Book Organization..3
    1.4.1   Overview..3
    1.4.2   Big Data Application and Analytics in a Large-Scale
            Power System..4
    1.4.3   The Role of Big Data Analytics in Smart Grid
            Communications... 4
    1.4.4   Big Data Optimization in Electric Power Systems: A Review 4
    1.4.5   Security Methods for Critical Infrastructure
            Communications 5
    1.4.6   Data-Mining Methods for Electricity Theft Detection................5
    1.4.7   Unit Commitment Control of Smart Grids 5
    1.4.8   Data-Based Transformer Differential Protection 5
1.5   Conclusions...6
References...6

1.1   Introduction

As a concept, big data and power systems might appear unrelated; however, the Smart Grid and advances in general computing power have made power systems a data-driven industry. The result of the ability to collect endless data is the emergence of big data. However, power systems are connected to physical devices and critical infrastructure (CI) and thus additional research problems and concerns exist in power system big data.

1.2 Big Data

Big data involves more than the size of the data itself and extends to the complexity and speed at which it is collected. The term big data is frequently defined with vague and self-referencing definitions and naturally big data logically extends from data (Bihl, Young II, & Weckman, 2016). While data are generally any sensed output, big data involves data that are too big, complex, or overwhelming to be analyzed by traditional methods (Bihl, Young II, & Weckman, 2016).

The primary attributes of big data are the 3 "V's" of *volume, variety,* and *velocity* (Bihl, Young II, & Weckman, 2016). While more than 42 attributes have been defined by some researchers in describing big data, the 3 V's capture the gist of the big data problem (see Shafer, 2017). As attributes, *volume* relates to the overall size of the data, *variety* indicates that big data can contain various types of data (text, strings, numbers, etc.) all within one dataset, and *velocity* indicates that big data is collected in real time (Bihl, Young II, & Weckman, 2016).

Critically, *velocity* is an attribute frequently associated with big data. Given enough time, any large volume and highly various dataset could eventually be analyzed using traditional methods. However, when these data are continuously being collected, a *velocity* problem exists whereby the growing size and complexity preclude traditional methods. Thus, advanced analytics and data management methods are both necessary (cf. Gutierrez, Boehmke, Bauer, Saie, & Bihl, 2018; Najafabadi et al., 2015).

1.3 Future Power Systems

Future power systems imply power systems that differ from today's due to increased decentralization, expanded communication and monitoring abilities, and wider variety of sources (Hebner, 2017). Multiple thrusts exist in power system research to accommodate this future; these include expanding the Smart Grid, increasing penetration of the Internet of Things (IoT), expanding renewable sources, and microgrid considerations.

Expanding penetration of the Smart Grid is not only expected but already underway (Amin & Wollenberg, 2005). Along with the Smart Grid comes a multitude of logged customer and power grid data which can be analyzed to find power theft (Jiang et al., 2014) and improve operating conditions of the grid at large (Fan et al., 2013). The IoT further expands upon the Smart Grid by enabling communication with any and all devices (Gubbi, Buyya, Marusic, & Palaniswami, 2013). An IoT-enabled power grid thus allows the monitoring of the CI while posing both big data and security problems (Sajid, Abbas, & Saleem, 2016).

Increasing decentralization through more microgrids and nanogrids can be also expected in the future power grid. While these have the ability to provide local resiliency (Hebner, 2017), they introduce uncertainty in larger grid planning (Khodaei, Bahramirad, & Shahidehpour, 2015). Added to this is the expected increase in the use of renewables, which also increase power system planning problems due to their general availability uncertainty (Atwa, El-Saadany, Salama, & Seethapathy, 2010; Polatidis, Haralambopoulos, Munda, & Vreeker, 2006).

1.4 Book Organization

To examine these problems, this book examines various intersections of big data and future power systems. For this goal, this book provides nine chapters, including the introduction, which focuses on the primary themes of big data in future power systems. Overall, this book discusses big data analysis methods, big data problems in future power systems, IoT concerns, security concerns related to big data, and various associated complexities.

1.4.1 Overview

This book is organized as follows:

- Chapter 2 discusses analytics and machine-learning methods in general and those applicable to big data in power systems.
- Chapter 3 discusses additional big data analytics relative to Smart Grid components.
- Chapter 4 discusses optimization methods which are suitable for big data models in power systems.
- Chapter 5 extends the discussion of Chapter 4 by considering various cyber security issues that exist in IoT-enabled future power systems.
- Chapter 6 discusses electricity theft detection and mitigation which is enabled by big data collection from the Smart Grid.
- Chapter 7 discusses renewable energy planning concerns which are associated with planned future power systems that have high renewable penetration.
- Chapter 8 discusses transformer protection methods which are enabled by big data collection on transformers.

1.4.2 Big Data Application and Analytics in
a Large-Scale Power System

To analyze big data, a variety of machine-learning methods are generally employed. Machine learning is broadly synonymous with pattern recognition, statistics, and data mining (Hand, 1998; Mannila, 1996). However, due to the emergence of big data, a variety of new methods have recently emerged, e.g., large-scale neural network known as "deep learning," which are capable to analyze and exploit the bigness of big data. While these methods have achieved significant advancements in image recognition, they have begun to see use in power system big data analysis (see LeCun, Bengio, & Hinton, 2015).

1.4.3 The Role of Big Data Analytics in
Smart Grid Communications

Because a Smart Grid can be described as a huge sensor network, with a lot of intelligent devices, the growth in the number of devices will produce a considerable amount of measured data. How to quantify and to analyze these data to enhance grid operation arises as one big concern. Advances of the Smart Grid promise to give operators and utilities a better understanding of customer behavior, demand consumption, weather forecast, power outages, and failures. However, it is vital to quantify the volume of sampled data to take advantage of them. Therefore, this chapter aims to characterize and to evaluate the emerging growth of data in communications network applied to Smart Grid scenario. A future active distribution system will serve as an example to demonstrate the data requirements for monitoring and controlling the grid.

1.4.4 Big Data Optimization in Electric
Power Systems: A Review

Traditional data-processing applications have difficulties operating effectively due to the complexity, velocity, and voluminosity of big data. This chapter presents a review of big data optimization problems in electric power systems. The chapter starts with scientometric mapping methods that show the variety and diversity of large-scale optimization problems in today's power system networks. An electrical grid power system could be categorized into generators which provide the required electric power, transmission systems that carry the electricity from the generating units, and distribution systems that feed the power to nearby industries and homes. The optimization issues such as logistics optimization in power system, as well as some optimization techniques including non-smooth, nonconvex, and unconstrained large-scale optimization are presented. Additionally, some metaheuristic methods in large-scale power system optimization problems have been reviewed.

1.4.5 Security Methods for Critical Infrastructure Communications

The proliferation of communication devices in CI applications presents security challenges. A variety of security approaches have been used to prevent unauthorized access to CI networks. This chapter will review (1) the communication devices used in CI, especially power systems, (2) security methods available to vet the identity of devices, and (3) general security threats in CI networks. Device identity verification methods will be discussed and range from bit-level, e.g., encryption keys, to physical layer, e.g., radio-frequency fingerprinting methods.

1.4.6 Data-Mining Methods for Electricity Theft Detection

Electricity theft is a major concern for utilities in both the developed and developing world. Although the United States has a low electricity theft rate, an estimated $4 billion of revenue is lost per year in the United States alone; the developing world generally sees much higher losses. Detecting potential electricity thieves is thus of interest to mitigate losses. Check meters and usage analysis have been used primarily to identify possible electricity thieves. However, advances in computing, the Smart Grid, smart meters, and in data mining have enabled more analysis to be conducted in this area. This chapter will review the wide variety of techniques and applications developed for electricity theft detection.

1.4.7 Unit Commitment Control of Smart Grids

Future power grids are planned to have significant renewable energy penetration. However, these sources of energy are unpredictable in nature. The unit commitment (UC) problem is the problem of producing power by collaboration of sources in order to achieve demand. This chapter discusses and presents a centralized approach to solve the UC problem for energy systems that contain a variety of generating components (traditional to renewable).

1.4.8 Data-Based Transformer Differential Protection

This chapter uses pattern recognition and dimensionality reduction methods for differential protection of power transformers. Both the linear principal component analysis (PCA) and the nonlinear, and neural network-based, curvilinear component analysis (CCA) are considered. Both PCA and CCA use the differential current from current transformers at transformer terminals. By using two techniques, this chapter illustrates how pattern recognition methods can be used to preprocess differential current to discernment internal faults currents (transformer differential protection zone) from inrush and over-excitation currents. Both PCA and CCA are employed with the Power System Computer Aided Design (PSCAD) electromagnetic simulation software in a three-phase power system, for distinct scenarios. The results show the feasibility to develop a differential protection to power transformers using data pattern recognition algorithms.

1.5 Conclusions

Overall, a wide variety of challenges exist in the future power grid, ranging from cyber security to data handling to planning. This book aims to discuss and present a variety of approaches to handling each of these challenges, in addition to discussions and reviews of the various topics and domains. To this aim, each chapter focuses on one specific topic and minimal overlap exists between chapters. However, the underlying theme in all chapters is the analysis and interpretation of big data due to future power system infrastructure.

References

Amin, S., & Wollenberg, B. (2005). Toward a smart grid: Power delivery for the 21st century. *IEEE Power and Energy Magazine, 3*(5), 34–41.

Atwa, Y., El-Saadany, E., Salama, M., & Seethapathy, R. (2010). Optimal renewable resources mix for distribution system energy loss minimization. *IEEE Transactions on Power Systems, 25*(1), 360–370.

Bihl, T., Young II, W., & Weckman, G. (2016). Defining, understanding, and addressing big data. *International Journal of Business Analytics (IJBAN), 3*(2), 1–32.

Fan, Z., Kulkarni, P., Gormus, S., Efthymiou, C., Kalogridis, G., Sooriyabandara, M., & Chin, W. (2013). Smart grid communications: Overview of research challenges, solutions, and standardization activities. *IEEE Communications Surveys & Tutorials, 15*(1), 21–38.

Gubbi, J., Buyya, R., Marusic, S., & Palaniswami, M. (2013). Internet of Things (IoT): A vision, architectural elements, and future directions. *Future Generation Computer Systems, 29*(7), 1645–1660.

Gutierrez, R., Boehmke, B., Bauer, K., Saie, C., & Bihl, T. J. (2018). Cyber anomaly detection: Using tabulated vectors and embedded analytics for efficient data mining. *Journal of Algorithms and Computational Technology*.

Hand, D. (1998). Data mining: Statistics and more? *The American Statistician, 52*(2), 112–118.

Hebner, R. (2017). The power grid in 2030. *IEEE Spectrum, 54*(4), 50–55.

Jiang, R., Lu, R., Wang, Y., Luo, J., Shen, C., & Shen, X. (2014). Energy-theft detection issues for advanced metering infrastructure in smart grid. *Tsinghua Science and Technology, 19*(2), 105–120.

Khodaei, A., Bahramirad, S., & Shahidehpour, M. (2015). Microgrid planning under uncertainty. *IEEE Transactions on Power Systems, 30*(5), 2417–2425.

LeCun, Y., Bengio, Y., & Hinton, G. (2015). Deep learning. *Nature, 521*(7553), 436.

Mannila, H. (1996). Data mining: Machine learning, statistics, and databases. *Eighth International Conference on Scientific and Statistical Database Systems*, Stockholm, Sweden, 2–9.

Najafabadi, M., Villanustre, F., Khoshgoftaar, T., Seliya, N., Wald, R., & Muharemagic, E. (2015). Deep learning applications and challenges in big data analytics. *Journal of Big Data, 2*(1), 1–21.

Polatidis, H., Haralambopoulos, D., Munda, G., & Vreeker, R. (2006). Selecting an appropriate multi-criteria decision analysis technique for renewable energy planning. *Energy Sources, Part B, 1*(2), 181–193.

Sajid, A., Abbas, H., & Saleem, K. (2016). Cloud-assisted IoT-based SCADA systems security: A review of the state of the art and future challenges. *IEEE Access, 4*, 1375–1384.

Shafer, T. (2017, April). *The 42 V's of big data and data science.* Retrieved from KD Nuggets. www.kdnuggets.com/2017/04/42-vs-big-data-data-science.html.

2

Big Data Application and Analytics in a Large-Scale Power System

Jeremy Lin
Transmission Analytics

Elham Foruzan
University of Nebraska-Lincoln

Fernando H. Magnago
Universidad de Rio Cuarto, Nexant Inc.

CONTENTS

2.1 Introduction 10
2.2 General Applications of Big Data 10
    2.2.1 Health Care 11
    2.2.2 Social Networking 12
    2.2.3 Handling Big Data 13
2.3 Algorithms for Processing Big Data...13
    2.3.1 Machine Learning and Deep Learning Generalities.................13
    2.3.2 Machine Learning 14
        2.3.2.1 Artificial Neural Network (ANN) Model.....................14
        2.3.2.2 Support Vector Machine (SVM) 15
        2.3.2.3 Decision-Tree Classifier...17
    2.3.3 Deep Learning...19
        2.3.3.1 Deep Learning Models..20
        2.3.3.2 Challenges and Suggested Solutions for Using
                Deep Learning in Big Data Analytics 25
2.4 Application of Big Data in Power Systems 27
    2.4.1 Big Data in Smart Grid Networks 27
    2.4.2 Phasor Measurement Units (PMU) 28
    2.4.3 Renewable Energy 29
    2.4.4 CIM as Information Standard for Big Data Analytics...............29
    2.4.5 Big Data Problem in Power System Modeling...........................30
        2.4.5.1 Security-Constrained Unit Commitment (SCUC)........30
        2.4.5.2 Decomposition Methods to Handle Big Data 31

2.4.5.3   Firm Transmission Right (FTR) Problems......................32
2.4.5.4   Time-Constrained Economic Dispatch...........................33
2.5   Conclusions...33
References..34

2.1 Introduction

Data are everywhere nowadays coming from an infinite number of sources. Images, videos, and encrypted data are all part of big data whose structure has become much more complex. Due to the high volume, velocity, and variety of data, this new breed of data is called "Big Data." Big data is a term that describes a large volume of data—both structured and unstructured—that inundates businesses, organizations, and lives on daily basis. But it is not the amount of data that is important. What really matters is what businesses and organizations do with the data. Big data can be analyzed for insights that lead to better decisions and strategic business moves. In the literature, the requirement to handle big data is known as "the 4Vs data." This 4Vs represent the characteristics of the data—volume, variety, velocity, and veracity.

With this big data, it is important to store it, process it, and be able to extract the value from that data. There, it becomes much more complicated and complex. Not only is there more data with more information that varies much more greatly, but now users also expect to do more with it. And they not only expect to do valuable things with their data, but also expect to extrapolate information and share data with other users' data. Collection, storage, management, and automated large-scale analysis of data are important functions to big data. The fundamental challenge of big data is not about collecting data, but about making sense out of it. The key questions related to big data are: What is the starting point? What are the computational paths to discovery of meaningful results? What are the relevant algorithms and how to visualize the findings? And what kind of key decisions can be made in the context of the application of big data? While the focus of this chapter is on the application and analysis of big data in a large-scale power system, we start with the general applications of big data in the next section.

2.2 General Applications of Big Data

"Big Data" demand cost-effective, innovative forms of information processing for enhanced insight and decision making. Explosive growth of big data was triggered by widespread adoption of the Internet around the globe. The Internet is essentially a realization of a concept of wide area networking based on computer and communication systems. The Internet is at once a world-wide broadcasting capability, a mechanism for information dissemination, and a medium for collaboration and interaction among individuals and their computers regardless of their geographic locations. Metaphorically, the Internet is like a gigantic information infrastructure as more and more institutions and individuals have joined to use the Internet as part of their daily routines.

With the explosive use of the Internet came the explosive amount of data. Huge amount of data are being collected and stored every day by many organizations. For example, Google typically processes over 20 petabytes (1 petabyte = 1,000,000,000,000,000 bytes) a day of user-generated data. Sources of these data include web data, e-commerce, purchases at department/grocery stores, bank/credit card transactions, social networks to name just a few.

The big data explosion touches every possible business and industry throughout the world, see (Bihl et al. 2016). Before discussing big data in power systems, two examples will be discussed to show the wide scale impact of big data. Both health care and social networking are used as two prime areas of big data impact and will serve as illustrative examples.

2.2.1 Health Care

In health care, there are many possible sources of big data. According to Gartner (Gartner March 2016) report, there are eight sources of big data in health care:

1. Physicians' free-text notes
2. Patient-generated health data (PGHD)
3. Genomics
4. Physiological monitoring data
5. Publicly available data
6. Credit card and purchasing data
7. Social media data
8. Medical imaging data

The volume of data for each source mentioned above can vary from approximately 100 GB per patient on genomics to terabytes of stored text (physicians' notes) to petabytes (medical imaging data). The structure of these data can range from free-hand/unstructured (physicians' notes) to some standard formats (genomics).

Health care analytics is also growing in importance, due to heath industry stakeholders' thirst for information, the need to manage large and diverse data sets, increased competition and growing regulatory complexity.

Innovations ranging from precision medicine to value-based care to population health management are also driving forces behind this. Value-based care relies on the foundation of robust data and analytics. The shift in the US health care system to value-based care is likely to demand significant analytic infrastructure investments and expansions across all health industry stakeholders interested in fully realizing and optimizing its value: health care providers and health systems, plans and payers, life sciences, and biopharma.

Despite this massive amount of data, it is important to improve the quality of information available to stakeholders in the health care system. Establishment of analytics program, information management, data governance, and IT platform are key to achieving quality improvement of information. As more and more data become available from sources like electronic health records, claims, wearable medical devices, social media, and the patients themselves, analytics can increasingly help detect patterns in information, delivering actionable insights, and enabling self-learning systems to predict, infer, and conceive alternatives that might not otherwise be obvious. In the future, such analytics-driven insights are likely to play a major role in helping health organizations reduce costs and improve quality, identify and better treat at-risk populations, connect with consumers, and better understand the performance and impact of health interventions on health outcomes.

The key challenge in health care business is to measure and improve the clinical performance to ensure clinical quality of the care delivered to the patients. Exploitation of full potential of clinical data is important to assess clinical quality. Clinical quality is equivalent to excellent clinical outcome, such as mortality, infections, survival curves, etc., and subsequent best path for patient treatment.

The raw data are not useful unless and until this siloed data can be transformed into patient-centered structure. The digital data should be easily accessible to patients and relevant stakeholders as well. Highly motivated physicians are willing and eager to adjust their clinical practice if presented with credible, high-quality data. Real-time information and predictive models should also be used to solve operational and clinical delivery problems. The key goal of using big data analytics is to find out which data source can significantly improve the analytics effort to help solve the problems in health care.

2.2.2 Social Networking

Most of the social networking are online in nature while relying largely on the Internet. With regards to data in social networking, there has been an explosive growth in size, complexity, and unstructured data. Significant research work has been done on big data in social networking which are enabled by various experimental methods including observational studies, and simulations, using large amount of data.

It is indeed "big data" which is the vast sets of information gathered by researchers at companies such as Facebook, Google, and Microsoft from patterns of cellphone calls, text messages, and Internet clicks by millions of worldwide users. Companies often refuse to make such information public, in part for competitive reasons and in part to protect customers' privacy.

2.2.3 Handling Big Data

There are two dimensions to enable the possibility of such big data: hardware capability and applications/algorithms. Hardware capability is comprised of storage capacity, network bandwidth, significantly increasing capability at the same cost or lower cost, and processing capacity. The main developments in applications/algorithms include online social networking, algorithmic breakthroughs, machine learning and data mining, cloud computing and its lower cost and scalability improvements, and ubiquitous sensors in every possible measurable point imaginable. For example, the price of 1 GB of storage has declined from roughly $300,000 in 1981 to $1000 in 1994 to a few cents nowadays.

2.3 Algorithms for Processing Big Data

Among the available algorithms used to process and analyze big data, machine learning and deep learning algorithms have gained significant attention recently in the research community. Although deep learning is a subset of machine learning, we will discuss them separately since deep learning builds upon methods and architectures found in machine learning.

2.3.1 Machine Learning and Deep Learning Generalities

The basic concept behind machine learning is to use unclassified training data to teach a machine through learning algorithms to inferred function in which the machine is able to classify new unseen data. Thus, machine-learning algorithms define weighing parameters for the models with the help of the training data. The weighing parameters are updated through finite iterations until the algorithm converges and learns to find patterns within the training data. After completing the training phase, the model can be used to predict the outcome of a variable if the new data are subsequently provided.

Generally, there are two types of machine learning: supervised learning and unsupervised learning. In the supervised learning approach, the algorithm is trained with data that contain both the input as well as the related output which are called labels. With the known output labels in the training data, the algorithm aims to determine a rule that maps attributes of input data to those labels. In the case of unsupervised learning, the training data does not contain any related output. Consequently, the algorithm involved cannot learn how to classify or predict new data points. Instead, unsupervised learning commonly aims at identifying a structure within the data.

2.3.2 Machine Learning

Machine learning provides different models that are capable in dealing with large and complex data sets. Additionally, machine learning offers different classification and prediction algorithms that can perform the predictions in the shortest time possible to facilitate real-time decision making in power system applications such as power market and system stability areas. Indeed, machine learning-based classification is also quite often applied in electricity markets, customer load prediction, and power system state estimation (Aggarwal et al. 2009; Saini et al. 2010; Soares et al. 2012).

We thus attempted to limit this chapter largely to some classification algorithms of machine learning, as we believe that this field has potentially important application in power systems. In this regard, various techniques such as fuzzy inference, fuzzy-neural models, artificial neural network (ANN), decision tree, and support vector machines (SVMs) (Deepak & Swarup 2011; Negnevitsky et al. 2009) are used in power system prediction and classification problems. Among the different AI methods, ANN, SVM, and decision-tree methods have received significant attention in recent years due to their high potential applications in power systems. In the rest of this section, three well-known supervised learning algorithms are further explained in detail. These three algorithms are ANN, SVM, and decision tree which are widely used for data analysis and classifications in power system problems.

2.3.2.1 Artificial Neural Network (ANN) Model

Neural networks have a considerable ability to obtain meaning from complicated data. It can be used for detecting and extracting the patterns from trends that are too complex to be noticed by either human or other computer techniques. ANN represents a structure of layers and interconnected processing nodes. Multi-layer perceptron (MLP) is the most widely used network architecture in ANN.

Figure 2.1 shows the basic scheme of a multilayer feed-forward ANN with three layers, namely input, hidden, and output layers. The circular nodes in Figure 2.1 represent the artificial neurons, while lines which are assigned with some weights represent connection from the output of one artificial neuron to the input of another artificial neuron (Kalogirou 2000). The input layer obtains input from its environment and sends it to the hidden layer. The purpose of the hidden layer is to connect the input layer to the output layer to extract more information for classification. The response of neurons is delivered by the output layer.

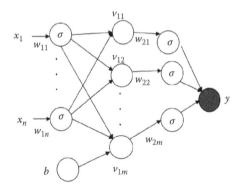

FIGURE 2.1
Scheme of a feed-forward ANN.

As can be seen in Figure 2.1, the input layer is vector $X = \{x_1, x_2, .., x_n\}$ and the output is y. The neuron k in jth hidden layer can be described as follows:

$$v_{jk} = \sigma\left(\sum_{i=0}^{n}(w_{ji}x_i + b)\right). \tag{2.1}$$

where σ (.) is the activation function, $w_j = \{w_{j1}, w_{j2}, ..., w_{jn}\}$ is the weight vector associated with input vector, and b is a bias level.

The goal of the ANN learning algorithm is to determine a set of weights, $W = \{w_1, w_2\}$, such that the sum of squared errors for the training data is minimized. Usually, the cost of the error, E, is calculated as the difference between the actual output and the desired output as

$$E = \left[y - f(W, X)\right]^2 \tag{2.2}$$

where $f(W, X)$ is the output determined by ANN.

2.3.2.2 Support Vector Machine (SVM)

Similar to ANNs, SVM is an algorithm that attempts to identify a function mapping that uses a number of input data from the training set and splits observations into several separate classes. Here, the general approach is to map all data into a new space in which data are separable, then find the best linear classification in the new space. The regression SVM problem can be stated as below.

Given the training data set with size n, (X_i, y_i) for $i = 1, ..., n$; in which X_i is the ith input vector, and y_i is the ith output vector, the target is to find the function $y(x)$ that can approximate the relation between input features and be able to predict output for the new input X (Alpaydin 2014). Using function

$g_j(\mathbf{x}), j = 1,...,m,$ input X is mapped into the m-dimensional feature space. Function $y(X)$ as the linear function of inputs in the new space is expressed as follows:

$$y(\mathbf{x},\omega) = \sum_{j=1}^{m} \omega_j g_j(\mathbf{x}) + b. \tag{2.3}$$

where w_j is the weight of input $g_j(\mathbf{x})$ and b is the bias term. SVM regression solves the problem to estimate the parameters $w_j, j = 1,...,m$ and the bias term. In the SVM, the ε-insensitive loss function is considered as an error. Therefore, error value less than ε is acceptable:

$$e_\varepsilon(r, y(\mathbf{x}, \omega)) = \begin{cases} 0 & \text{if } |r - y(\mathbf{x}, \omega)| \le \varepsilon \\ |r - y(\mathbf{x}, \omega)| - \varepsilon & \text{otherwise} \end{cases} \tag{2.4}$$

Thus, SVM regression is formulated to minimize the error function and problem complexity:

$$\min \frac{1}{2} \| \omega \|^2 + C \sum_{i=1}^{n} (\xi_i + \xi_i^*).$$

$$\text{s.t.} \begin{cases} r_i - y(\mathbf{x}_i, \omega) \le \varepsilon + \xi_i^* \\ y(\mathbf{x}_i, \omega) - r_i \le \varepsilon + \xi_i \\ \xi_i, \xi_i^* \ge 0, i = 1,...,n \end{cases} \tag{2.5}$$

where ξ_i^* and ξ_i are the upper and lower training errors, respectively. The optimization problem in Equation (2.5) can be transformed into a dual problem, and its solution is obtained by maximizing the dual function as shown in the following equation:

$$y(\mathbf{x}) = -1/2 \sum_t \sum_s |(\alpha_+^t - \alpha_-^t)(\alpha_+^s - \alpha_-^s) K(\mathbf{x}_i, \mathbf{x}) - \varepsilon \sum_s (\alpha_+^t + \alpha_-^t) - \sum_s r^t (\alpha_+^t - \alpha_-^t).$$

$$\tag{2.6}$$

$$\text{s.t. } 0 \le \alpha_+^t \le C, \quad 0 \le \alpha_-^t \le C, \sum_t (\alpha_+^t - \alpha_-^t) = 0$$

where $K(\mathbf{x}, \mathbf{x}_i)$ is defined as follows:

$$K(\mathbf{x}, \mathbf{x}_i) = \sum_{j=1}^{m} g_j(\mathbf{x}) g_j(\mathbf{x}_i). \tag{2.7}$$

Parameter C determines the trade-off between the model complexity and the degree of acceptable error. Increasing the amount of C will increase the effect of minimizing error. It is proved that we can estimate $K(\mathbf{x}, \mathbf{x}_i)$ with a kernel function. The common kernel function that is applied for SVM is RBF kernel which is shown as follows:

$$K(\mathbf{x}_i, \mathbf{x}) = \exp\left(-\frac{\|\mathbf{x} - \mathbf{x}_i\|^2}{2p^2}\right) \tag{2.8}$$

By solving Equation (2.6), coefficients α_+^t, α_-^t will be calculated. Finally, function $y(\mathbf{x})$ can be written as a weighted sum of the support vectors

$$y(\mathbf{x}) = \sum_{i=1}^{n_{SV}} (\alpha_i - \alpha_i^*) K(\mathbf{x}_i, \mathbf{x}) + b. \tag{2.9}$$

2.3.2.3 Decision-Tree Classifier

The decision-tree classifier is one of the possible approaches to multi-stage decision making, which uses the recursive top–down approach of decision-tree structure. A decision tree has a tree structure starting with a root node that is connected to the internal nodes using tree edges. The internal nodes recursively partition the instance space into two or more subclasses until tree leaf nodes are reached. Each leaf node is assigned to one class that is the most appropriate among all the classes. Therefore, a decision tree consists of a root node with no incoming edges, leaf nodes with no outgoing edges, and the internal nodes that have only one incoming edge.

There are many algorithms that can be used to determine the best way to partition the data in each internal node and build a final decision tree. Among them, the CART algorithm is a classification and regression tree that fits well with our numeric data space (Rutkowski et al. 2014). In this section, we describe the CART algorithm that can classify the real value attributes into two classes. The resulting full-grown tree is identical to the tree constructed by the algorithm.

a. The CART Algorithm

Gini gain is generally used to determine the suitable attribute to partition the data in root and each internal node. Therefore, for each node i, the attribute with the highest Gini gain ($G(.)$) is selected to partition the data set coming from the parent node to the node i. In the binary CART, the algorithms recursively divide every node into left and right partitions (Roman 2004; Rutkowski et al. 2014). The node partitioning continues until a stopping criterion is triggered. The concepts of Gini index and Gini gain are further described below. The stopping criteria can be defined as setting the maximum tree depth, or if all data are classified with a deep tree.

b. Gini Index

The Gini index is used to measure impurity of each node in the CART algorithm. Let's suppose that node i processes data set S_i that comes from node i's parent. The Gini index at node i is calculated from the following expression (Roman 2004):

$$\text{Gini}(S_i) = 1 - \sum_{k=1}^{K}(F_{k,i})^2. \tag{2.10}$$

where $F_{k,i}$ is the fraction of all the data in S_i that belongs to class $k \in \{1, 2, ..., K\}$. The minimum value of Gini index is obtained when all data are coming from one class. And the maximum value of Gini is obtained when the data are equally distributed among all classes.

c. Gini Gain

For each attribute j selected from the set of N available attributes, $j \in \{1, 2, ..., N\}$, the set of attribute values A^j is partitioned into two disjoint subsets A_L^j and A_R^j. The two subsets A_L^j and A_R^j are complementary and their union is set A^j. Suppose that P_i represents the set of all possible partitioning subsets of set A^j. Every possible partitioning from set P_i results into different subsets A_L^j and A_R^j and divides the dataset S_i at node i, into two disjoint left and right subsets, $L_i(A_L^j, A_R^j)$ and $R_i(A_L^j, A_R^j)$. Now, if the $F_{L,i}(A_L^j, A_R^j)$ and $F_{R,i}(A_L^j, A_R^j)$ represent the fractions of data element from S_i belonging to $L_i(A_L^j, A_R^j)$ and $R_i(A_L^j, A_R^j)$, respectively, we have

$$F_{L,i}(A_L^j, A_R^j) + F_{R,i}(A_L^j, A_R^j) = 1 \tag{2.11}$$

The fraction of data from $L_i(A_L^j, A_R^j)$ and $R_i(A_L^j, A_R^j)$ from class $k \in \{1, 2, ..., K\}$ are denoted with $F_{L,i,k}(A_L^j, A_R^j)$ and $F_{R,i,k}(A_L^j, A_R^j)$. The weighted Gini index of set S_i with partition sets A_L^j and A_R^j is defined as follows:

$$\text{Weighted\_Gini}(S_i, A_L^j, A_R^j) = F_{L,i}(A_L^j, A_R^j)\text{Gini}(L_i(A_L^j, A_R^j))$$

$$+ F_{R,i}(A_L^j, A_R^j)\,\text{Gini}(R_i(A_L^j, A_R^j))$$

where

$$\text{Gini}(L_i(A_L^j, A_R^j)) = 1 - \sum_{k=1}^{K}(F_{L,i,k}(A_L^j, A_R^j))^2. \tag{2.12}$$

$$\text{Gini}\left(R_i\left(A_L^j, A_R^j\right)\right) = 1 - \sum_{k=1}^{K}\left(F_{R, i, k}\left(A_L^j, A_R^j\right)\right)^2. \tag{2.13}$$

Finally, the value of Gini gain is calculated as follows:

$$G\left(S_i, A_L^j, A_R^j\right) = \text{Gini}\left(S_i\right) - \text{Weighted}_{\text{Gini}\left(S_i, A_L^j, A_R^j\right)}. \tag{2.14}$$

Among all possible partitions of set P_i, the partition which maximizes the value of Gini gain is chosen as an optimal partition of set A_j for the subset of data S_i:

$$A_j^* = \arg\max_{A_L^j, A_R^j \in P_i}\left\{G\left(S_i, A_L^j, A_R^j\right)\right\}. \tag{2.15}$$

2.3.3 Deep Learning

In the modern world of information technology and smart devices, tremendous amounts of data are created every day. On average, 2.5 quintillion data are created daily (Wu et al. 2014). Our capability to produce data is enormous in the current century since large-scale quantities of data such as digital streams of measurement in the form of text, image, and video for different purposes such as better monitoring or security are being collected and made available across various domains, including power systems. Therefore, there is a potential to efficiently use these data to improve the stability, robustness, and economics of the power system.

Nevertheless, the benefit of massive data and the presence of such an enormous data inevitably lead to the important challenge of dealing with big data. Big data refers to the exponential growth and wide availability of data that are difficult to store, process, manage, and analyze within a "tolerable elapsed time" using commonly used software tools and technologies (Zhi-Hua et al. 2014). In order to take advantage of available big data, it is necessary to develop tools and methods that can be applied to explore and extract useful information, patterns or knowledge from large-scale data. Meaningful information and patterns extracted from large-scale input data are used for future actions such as decision making and prediction, which are at the core of big data analytics. Big data analytics aims to develop novel algorithms and models to address specific issues related to big data.

Deep learning provides one such model for analyzing available big data. The complex abstractions and data representations from large volumes of data, especially unsupervised data, by deep learning can be considered as a practical source of knowledge for decision making, information retrieval, and for other purposes in big data analytics. Indeed, certain big data domains, such as computer vision (Krizhevsky et al. 2012) and speech recognition (Hinton et al. 2012), take advantage of deep learning to improve classification modeling results.

In this section, we will first introduce deep learning as a tool of big data analytics. Then, we will review three deep learning architectures that are most commonly used. Finally, we will discuss the challenges and some solutions of using deep learning in big data analysis.

2.3.3.1 Deep Learning Models

Deep learning algorithms are represented by architectures of consecutive layers. The objective of this deep architecture is to learn complicated representation of the data in a hierarchical manner by passing the data through multiple stacked layers. Each layer applies a nonlinear transformation on its input and provides automated feature selection to its output. In this architecture, the input data are fed to the first layer and then the output of each layer is provided as input to the next layer, consecutively.

The architecture of stacking up the nonlinear transformation layers is the main structure in deep learning algorithms that help extract different features from raw input data. With increasing number of layers in this architecture, more complicated nonlinear transformations can be constructed. These nonlinear transformation layers extract different features from input data and represent the data in different layers; so deep learning possesses structures with multiple levels of data representations. The achieved final representation using deep learning algorithms is a highly nonlinear function of the input data.

The main advantage of deep learning algorithms is that they automatically extract features from complex and massive data (Bengio 2009, 2013). Deep-learning algorithms do not attempt to construct a pre-defined sequence of representations at each layer as in the most machine learning algorithms, but instead perform nonlinear transformations in different layers. These transformations disentangle factors of variations in data in different layers. Therefore, deep learning not only provides complex representations (feature selection) of data but also makes the machines independent of human knowledge which is the ultimate goal of big data analysis. In this regard, deep learning models receive massive amount of unsupervised and supervised data as an input to automatically extract complex patterns in the input data. And these models extract available patterns directly from unsupervised and supervised data without human interference. Another advantage of deep learning algorithms is their ability to analyze unsupervised data, which makes them more suitable for large percentage of available data. Unsupervised learning process is intended to learn data distributions without using label information since data are largely unlabeled. On the other hand, the supervised data have labels for all available data.

    a. Convolutional Neural Network (CNN)

        A convolutional neural network (CNN) is a special case of the neural network which consists of one or more layers for feature representations (or feature maps) which are followed by one or more fully connected layers as in a standard neural network for classification (Hijazi et al. 2015; Krizhevsky et al. 2009). Layers for feature selection usually consist of two types of layers called convolutional and pooling/subsampling layers. Convolutional layers perform convolution operations with several filter maps of equal size, while subsampling layers reduce the sizes of proceeding layers by averaging data within a small neighborhood or max-pooling. Figure 2.2 illustrates a typical CNN network. The input is converted with a set of filters called feature maps. Then, pooling/subsampling layers are applied to reduce the dimensionality of filtered data. The number of layers for feature representation depends both on the problem complexity and the designer discretion.

Convolutional layers are an essential part of multi-layer CNN. The layers' parameters consist of a set of learnable filters, which extracts different features of the input. In a multilayer CNN, the first convolution layer extracts low level features while higher level layers extract higher level features (Krizhevsky et al. 2009). A fully connected layer is used for classification during the training process.

Each convolutional layer is composed of multiple feature maps, which are constructed by convolving inputs with different filters. Each filter is convolved across the input volume by computing the dot product between the filter and the input to detect local features

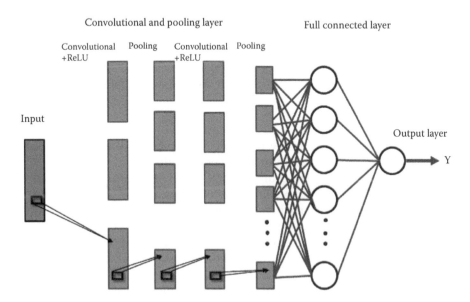

FIGURE 2.2
Scheme of a feed-forward CNN.

during the feed-forward process. Following each pooling layer is an element-wise nonlinearity, which allows the CNN to learn new kinds of nonlinearity (Hijazi et al. 2015; Krizhevsky et al. 2009). Mathematically, applying local filters and a nonlinear function is stated below. The value of a neuron vx_{ij} at position x of the jth feature map in the ith layer is calculated as follows:

$$v_{ij}^x = g\left(b_{ij} + \sum_m \sum_{p=0}^{P_i-1} \omega_{ijm}^x \ v_{(i-1)m}^{x+p} \right). \tag{2.16}$$

where m is the feature map in the $(i-1)$th layer connected to the current feature map; $\omega^p{}_{ijm}$ is the weight of position p connected to the mth feature map; P_i is the width of the kernel toward the spectral dimension; and b_{ij} is the bias of jth feature map in the ith layer; $g(.)$ is a nonlinear function which introduces nonlinearity into the model. *Relu* activation function is an option as a nonlinear function:

$$g(x) = \text{Relu}(x) = \max(0, \ x). \tag{2.17}$$

Each neuron on the convolutional layer is connected to a local region of the previous layer and shares weights with other neurons on the same feature map to control the network capacity. A pooling layer is usually inserted in between successive convolutional layers to reduce the spatial size of the representation for decreasing the number of parameters and computations in the network and for controlling over-fitting. Each pooling layer corresponds to the previous convolutional layer. The neuron in the pooling layer combines a small $N \times N$ patch of the convolution layer. The most common pooling operation is max pooling which is expressed as follows:

$$a_j = \max_{N \times N} \left(a_i^{n \times 1} u(n, 1) \right). \tag{2.18}$$

where $u(n, 1)$ is a window function to the patch of the convolution layer, and a_j is the maximum in the neighborhood.

At the end of the structure of CNN model, the feature map as the output of the last max-pooling layer is then fed into the penultimate fully-connected layer where the neurons are fully connected to all activations in the previous layer, same as regular neural network described. The fully-connected layers are capable of combining the features abstracted from lower layers for final classification.

The weights among all layers, including the convolutional layers and fully connected layer of the deep CNN model, are trained using a backpropagation algorithm and a gradient descent algorithm with mean squared-error as the loss function.

b. Recurrent Neural Network (RNN)

Recurrent neural networks (RNNs) contain cyclic connections that make them a more powerful tool to model sequence data. These models learn to map input sequences to output sequences via a continuous vector valued intermediate hidden state. RNNs contain cycles that feed previous time step results into the network as a current input to have predictions at the current time step. These results are stored in the intermediate states of the RNN network. Therefore, in contrast to other algorithms that are designed for static windows of input data, the RNN can capture dynamically changing contextual windows over the input using long short-term memory (LSTM) architecture that is designed in RNN algorithm. LSTM is capable of learning long-term dependencies within a sequence of data (Hasim et al. 2014). It contains special units called memory blocks in the recurrent hidden layer. The memory blocks contain memory cells with self-connections storing the temporal state of the network in addition to special multiplicative units called gates to control the flow of information. Each memory block in the architecture contains an input gate, forget gate, and an output gate. The role of each gate is explained as follows (Hasim et al. 2014):

1. **Input gate**: to control the flow of input activations into the memory cell.
2. **Output gate**: to control the output flow of cell activations into the rest of the network.
3. **Forget gate**: to scale the internal state of the cell before adding it as input to the cell through the self-recurrent connection of the cell, therefore adaptively forgetting or resetting the cell's memory.

Different structures of RNN are reported in the literature (Hasim et al. 2014). Figure 2.3 shows a standard structure of LSTM RNN architecture. This structure has an input layer, a recurrent LSTM layer and an output layer. The input layer is connected to the LSTM layer. The LSTM output units are also connected to the output layer of the network.

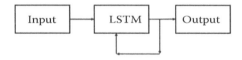

FIGURE 2.3
Standard LSTM RNN architecture.

c. Deep Belief Network (DBN)

Similar to CNN, deep belief network (DBN) has an input layer, stacks of hidden layers, and output layer. One of the big advantages of DBN is their capability of feature representation for both labeled (supervised) data and unlabeled (unsupervised) data (Hinton & Salakhutdinov 2006). Figure 2.4 shows a typical DBN architecture, which consists of restricted Boltzmann machines (RBMs) layers and/or one or more additional layers for discrimination tasks (Hinton & Salakhutdinov 2006). Each RBM layer consists of two consecutive layers of nodes, in which all the nodes from one layer are connected to all nodes in other layers. Similar to other deep learning algorithms, the final goal of DBN is to train weighing parameters in the network.

In this network, learning starts with unsupervised learning of each RBM using a Gibbs sampling and then updates the parameters for the RBM layer (Xue-Wen & Lin 2014). In this method, pre-training of RBMs is performed first and the output of each layer is fed to the next RBM layer. This pre-training is unsupervised, as unlabeled data are used for training RBMs. For each RMB with assumed Bernoulli distribution, the sampling probability is as follows:

$$p(h_j = 1 \mid v; W) = \sigma \left(a_j + \sum_I w_{ij} v_i \right). \tag{2.19}$$

$$p(v_i = 1 \mid h; W) = \sigma b. \tag{2.20}$$

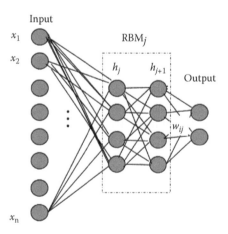

FIGURE 2.4
Schematic of DBN architecture.

where v represents $I \times 1$ input unit vector and h represents $J \times 1$ hidden unit vector; W is a matrix of weight connecting a hidden layer to its input layer (previous hidden layer); and $\sigma(.)$ is a sigmoid function. Weights w_{ij} are updated based on a contrastive divergence approximation and are shown as follows (Hinton 2002; Xue-Wen & Lin 2014):

$$\Delta w_{ij}(t+1) = c\Delta w_{ij} + \alpha\left(\left\langle v_i h_j \right\rangle_{\text{data}} - \left\langle v_i h_j \right\rangle_{\text{model}}\right). \tag{2.21}$$

where α is the learning rate and c is the momentum factor and $\langle . \rangle_{\text{data}}$ and $\langle . \rangle_{\text{model}}$ are the expectations of distribution for data and model.

Therefore, RBM training includes a Gibbs sampler to sample both hidden and its input layers and uses the contrastive divergence approximation to update the weights between these two layers. This process repeats several times until weights converge.

2.3.3.2 Challenges and Suggested Solutions for Using Deep Learning in Big Data Analytics

In this section, we provide three characteristics that are important parts of analyzing big data: (1) high volume, (2) high velocity, and (3) high variety (Xue-Wen & Lin 2014). These three characteristics refer to: (1) large scale of data, (2) high speed of streaming data, and (3) different types of data. Due to these characteristics, it is indeed challenging to develop deep learning algorithms.

a. Deep Learning for High-Volumes of Data

There are two big challenges associated with high-volumes of data. First, data that are used to build a deep learning algorithm usually possess large numbers of examples, high dimensionality of attributes, and varieties of output classes. These properties may lead to model complexity. Also these complex models can have a very long running time. In these cases, working with just one central processor is difficult or even impossible. The second challenge is associated with noisy labels. Data may have been collected from different sources during long periods of time. Therefore, data may be mislabeled or not labeled at all. Distributed programming with parallel machines is one possible solution to address the first challenge. Parallelizing several CPUs and GPUs increases the training speed without scarifying the model accuracy. Indeed, novel algorithms have used parallel processing to create deep learning models (Xue-Wen & Lin 2014).

The second challenge of high-volumes of data can be addressed with a natural ability of deep learning algorithms to extract features from unlabeled data. Thus, since most of the available data are unlabeled or noisy data, utilizing an advanced deep learning algorithm is a perfect solution to extract patterns available in big data. In this regard, some researchers have used semi-supervised learning to alleviate noisy data problems.

b. Deep Learning for High-velocity of Data

In today's data-intensive era, data velocity—the increasing rate at which data are collected and obtained—is another challenge for deep learning algorithms. Data can be produced at an extremely high speed and may need to be processed in a timely manner. One example of high-velocity data in power system is the PMU data which are usually collected with the frequency of 30–60 data samples per second.

Online learning is one possible solution for high-velocity data which is to learn one instance at a time and update network parameters. To speed up sequential online learning process, researchers performed the network update on mini-batch size of streaming input data (Scherer et al. 2010). Also, if a possibility of data loss with streaming data exists, and if it is generally not immediately processed and analyzed, there is an option to save fast-moving data into the bulk storage for processing them at later time. However, the high-velocity nature of big data is a challenging problem that needs further investigation.

c. Deep Learning for High Variety of Data

High variety of big data creates another challenge for deep learning algorithms. These days, data produced from different sources and combination of data coming from different sources form one complete data set. For example, data that need to be analyzed can be a collection of messages, images, and audio streams, each type of data coming from different probability distributions.

As mentioned before, a natural characteristic of deep learning method is their ability to classify supervised and unsupervised data, or a combination of both through the hierarchal learning process, in which each layer can capture different features of data. The abstract representations provided by deep learning algorithms can separate the different sources of variations in data. Therefore, one solution to address high-variety of data is to learn data representation (feature selection) from each data source individually, and then combine them in an appropriate deep learning structure. For example, authors in (Srivastava & Salakhutdinov 2012) developed a deep learning module that is aimed to find patterns from two different sources, image and text data. In their deep learning model, the authors first built two separate deep learning structures for image and text data. Then, the additional layer was developed to build the joint representation of all data.

2.4 Application of Big Data in Power Systems

During the last ten years, there has been a remarkable increase of data available in different areas of power systems, such as data for analyzing power market and data from time-synchronized phasor also known as phasor measurement units (PMUs) for state estimation. Therefore, it is necessary to extract insights from available data and enhance power quality and optimize power system operations. These new data that are rich with information can provide many insights and stimulate research opportunities for power system enhancement. For example, by utilizing the huge data that are available from different sources, power system companies and market participants can increase their performance and utilization of system assets. Optimal application of time-synchronized phasor data leads to the development of wide-area measurement systems (WAMS), recently. WAMS is a powerful approach to identify inter-area oscillations in a large grid. In these applications, PMU-based wide-area measurements are being used to provide remote feedback signals and improve the damping of inter-area oscillations through specially designed damping controllers (Raoufat et al. 2016). Other design techniques and methodologies have also been reported for PMU measurements including fault-tolerant (Raoufat et al. 2017; Foruzan et al. 2017) and coordinated damping controllers.

However, to extract necessary information and gain useful insights from available data to enhance power system, the classification and predictive models are useful and necessary to achieve that purpose. For example, to manage the risks in electricity markets, it is necessary to forecast different market indicators such as the hourly price of the spot markets, customer loads, and renewable energy productions. Fairly accurate forecasts of load, energy production, and market prices are important inputs to the decision-making activities of a generation company or an electric utility for producing energy. Electricity is a special commodity which is not easily storable. All generated electricity must be consumed in the instant it was produced. Therefore, both producers and consumers need accurate price forecasts to establish the best strategies for their own benefits.

2.4.1 Big Data in Smart Grid Networks

Nowadays, smart grid (SG) technology can fulfill the new requirements to manage a distribution system efficiently. This task is performed by incorporating advanced information and communications technology (ICT). The extensive deployment of the advanced ICT, materialized in part by smart meters, is producing significant amounts of data regarding memory, speed, and heterogeneity. The generated big data bring substantial benefits to help manage the system more efficiently. However, handling this amount of data presents several challenges. That is the reason why big data technology is a new scientific trend within the SG area.

In this scenario, the data processing is of primary concern and its urgency increases with data growth. For the particular case of SGs, traditional model-based tools need to be modified because the big data need to be handled within short elapsed time and using limited hardware resources. This new paradigm, known as big data technology, must be interpreted as an extension of the traditional methods (He et al. 2017).

2.4.2 Phasor Measurement Units (PMU)

Synchrophasors are time-synchronized vectors that represent both the magnitude and phase angle of the sine waves found in electricity, and are time-synchronized for accuracy. They are measured by high-speed monitors called PMUs that are about 100 times faster than the measurements provided by existing Supervisory Control and Data Acquisition (SCADA) system. PMUs measure current and voltage by amplitude and phase at selected locations of the transmission system. The high-precision time synchronization (via GPS) allows comparing measured values (synchrophasors) from different substations far apart and drawing conclusions as to the system state and dynamic events such as power swing conditions. PMU measurements record grid conditions with high accuracy and offer insight into grid stability or stress. Synchrophasor technology is used for real-time operations and off-line engineering analyses to improve grid reliability and efficiency and lower operating costs.

PMUs are typically used for wide-area monitoring and grid monitoring, as its measured variables use synchrophasors from PMUs serving as sensors. It helps with quick recognition of the current network situation and indicates both power swings and transient phenomena, transparently as well as instantly. Measurements from PMUs support control center personnel in assessing critical grid situations and contribute to taking suitable actions. As all measured results are stored, power system disturbances can be promptly analyzed. The PMU devices determine current and voltage phasors with highly accurate time stamps and transmit them for analysis together with other measured values (frequency, speed of frequency change) using the IEEE C37.118 communication protocol, which are typically sent to the control centers. PMUs and their measured synchrophasors make a valuable contribution to the dynamic monitoring of transient processes in energy supply systems.

System frequency is one of the electric grid's "vital signs" much the same as a human's pulse or temperature. The University of Tennessee in Knoxville, in collaboration with Oak Ridge National Laboratory, deployed a system of global positioning system (GPS) synchronized sensors to measure the voltage angle and frequency of the electric grid on a wide-area basis. This is the largest wide-area electric grid sensor network in the world—allowing power system engineers to see the dynamic behavior of the total interconnected electric grid and to understand how the various geographical regions interact with one another. These data flow from the remote sensors into the central processing facility in Knoxville, Tennessee where it is time-synchronized and incorporated into the map. These data are used by experts across North America when they investigate electricity blackouts.

The North American SynchroPhasor Initiative (NASPI) is a collaborative effort among the U.S. Department of Energy (DOE), the electric power industry, and academia. NASPI was established in 2007 to advance the understanding and use of synchrophasor technology, and is a forum for Recovery Act Smart Grid Investment recipients to share information and lessons learned, solve problems, develop technical standards, and further the development of synchrophasor technology. As a result, much of the collective insights and work products produced by NASPI are direct outcomes of the Recovery Act Smart Grid Investments. DOE has funded NASPI research and national laboratory participation in NASPI since 2007, and began supporting NASPI directly at the start of 2014.

2.4.3 Renewable Energy

Neural networks have been seen as very useful in the area of renewable energy, neural networks models can be used to estimate output power as a function of wind turbine parameters and delay of corresponding parameters (i.e., power, wind speed). Humidity, wind speed, and time are used as input variables to train a neural network model in power prediction applications. Prediction of short-term and long-term power using the k-NN algorithm has also been proposed. Analysis results based on power estimation-based clustering method were also reported. These research works show the importance and challenge of extracting the correct information from big data sources, and producing a useful and reduced set of data. Nowadays, the major areas that apply DM methodology to handle big data are security assessment, fault detection, power system control, load forecasting, and load profiling (Hu & Vasilakos 2016).

2.4.4 CIM as Information Standard for Big Data Analytics

In addition to the increasing need for robust algorithms that can handle large amount of data, the interoperability issue between information systems provided by different software/hardware vendors containing information of very large networks is attracting more attention. As an example, the European Network of Transmission System Operators (ENTSO-E) has been conducting large-scale interoperability tests for grid model data exchanges since 2009. These tests provide a voluntary environment to ensure relatively easy and seamless data integration using Common Information Model (CIM) standard among different software applications for transmission network planning and operations. CIM was first initiated by Electric Power Research Institute (EPRI) more than 15 years ago. Since then, the International Electrotechnical Commission (IEC) Technical Committee 57 (TC 57) has been active in managing and expanding the CIM model. A series of standards released by IEC TC 57 started to form the base for network data design, exchange and transfer for power system applications within electric power utilities.

To address the data interoperability issue and facilitate data exchange among different applications, many utilities have standardized or are in the process of standardizing their data model exchange using CIM standard. Most of the SCADA systems, databases, and smart devices in operation support CIM XML standard for data exchange. To work with this scenario, the programs need to provide a built-in CIM data interface, import network data definitions in CIM XML format, render XML data into the user interfaces and use it as an input to the power system analysis algorithms. The challenge is on the approach to handle big data using this type of format (Magnago et al. 2015).

2.4.5 Big Data Problem in Power System Modeling

Big data issues also exist in power system modeling applied to different areas such as scheduling, unit commitment or electricity market tools. The real needs related to big data stem from emerging problems such as increasing numbers of constraints and more periods of time needed in optimization formulations, different network topology changes, and more bids/offers in electricity markets (Hong et al. 2016).

2.4.5.1 Security-Constrained Unit Commitment (SCUC)

To illustrate the potential enormity of the problem in power system security analysis, consider the following somewhat extreme short-term SCUC example. Let the SCUC problem have 24 hourly time periods, and assume that the outer loop in Figure 2.5 is cycled ten times to capture and resolve all constraint violations. For a network of 20,000 buses and a list of 1,000 contingencies, this SCUC calculation involves the solution of 240,000 contingency cases. Suppose that we use an AC network model throughout, where the time to obtain a solution for each post-contingency AC power flow takes about 0.25 s. Then with the simplest iteration strategy, the study's security analysis takes a total of 16.66 h on a single CPU. This of course does not take into account of the use of good critical constraint-set iteration strategies, which can easily reduce the overall time by a factor of five or more.

At the other extreme, if the simplest DC-type network model with LODFs (line outage distribution factors) is used throughout, the equivalent security analysis calculation is in the orders of magnitude faster. However, additional tools are needed to analyze and account for differences between AC and DC flows and AC model is required to deal with voltage issues that are ignored in DC models. In summary, the research needs to address the combination of

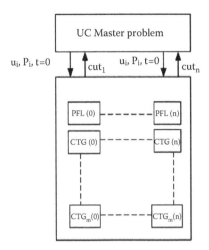

FIGURE 2.5
Security analysis: base case and contingencies.

large-scale problems with an increasing number of variables while providing a solution with high accuracy requirement (i.e., very small relative gap), and a limited solution time (i.e., for a day-ahead problem, the current time limit is 1200 s) (Pinto et al. 2006).

2.4.5.2 Decomposition Methods to Handle Big Data

There are different decomposition–coordination schemes that can tackle large-scale multi-stage problems of both deterministic and stochastic natures. Decomposition methods aim to reduce large-scale problems into simpler problems. These methods take advantage of the fact that although the problems in power systems are large, the problem structure can be decomposed into subproblems, such as decomposed problems for different periods, different contingencies, etc.

The first decomposition methodology that was proposed to solve these types of problems is the Dantzig–Wolfe decomposition method. The basic idea of this method is to build an equivalent master problem with fewer rows than the original problem but with a very large number of columns. This master problem can be solved using any linear programming (LP) technique. Then, the multiplier parameters or prices are sent to the subproblems. The subproblems are then solved and the results are sent back to the master problem which combines these results with the previous solution and calculates a new price. This process of looping between master problem and subproblems is iterated until an optimal solution is obtained. This technique works particularly well when the size of the problem is large in term of columns. Additional methods were proposed when the problem size increases in row direction. One of these techniques is known as the Benders Decomposition (BD) method.

BD is an algorithm that has been broadly used for large-scale optimization problems, particularly for those in power systems. BD has three main advantages: (1) modularity, (2) flexibility, and (3) robustness. With respect to modularity, master and subproblems can be separately solved by specialized algorithms, thus providing speed and efficiency on the overall performance of the global optimization process. Additionally, Benders flexibility is mainly supported by the different existing power system applications. For example, it is possible to find its application in areas like security-constrained economic dispatch (SCED), generation-transmission planning, hydrothermal coordination, and optimal power flow. Finally, in terms of robustness, despite the different natures of the master problems and the subproblems in SCUC applications, both problems are essentially solved using LP algorithms. This is an important feature because LP algorithms are one of the most mature and proven methodologies among other optimization techniques. Nevertheless, since BD is a cutting-plane method, it may present instabilities which are translated into delays in the algorithm convergence. In addition, since the master level is formulated as a mixed-integer linear problem (MILP), the convergence time is strongly affected by the high computational burden of the master problem. Therefore, there are many ongoing research efforts with the goal of improving BD performance. Among the different suggested possibilities, having a better initialization of BD is recognized by several researchers as one of the most important enhancements by concluding that it could have a significant effect on BD performance (Wang & Shahidehpour 1992).

There are two types of methodologies to solve stochastic problems: (1) Progressive Hedging (PH) method and (2) Dual Approximate Dynamic Programming algorithm (DADP). The PH method decomposes the problem into different scenarios and includes some of the constraints as a dual problem. The main advantage of this approach is that it can solve large numbers of dynamical subsystems while the disadvantage is that the complexity increases with the number of stages. The DADP decomposes the problem spatially and considers the dynamic subsystems as a dual problem. The complexity linearly increases with the number of stages, depending on the approximation supposed to solve the subproblems by the dynamic programming method (Gangammanavar et al. 2016).

2.4.5.3 Firm Transmission Right (FTR) Problems

Financial transmission rights (FTRs) play a crucial role in electricity market designs since these market products allow the market participants to hedge against highly varying market prices by reducing price uncertainty as well as facilitating competitive open transmission access. FTRs, along with other market products, have been used since 1998 in many well-known electricity markets in the United States. The evolution of these markets shows that the number of pricing nodes and market activities increased considerably while the virtual transaction volumes tripled in the past few years. Additionally, the evolution of the FTR analysis also includes an increasing number of contingencies. Moreover, there are other FTR markets that consider the multi-period case which increases the problem not only in the direction of the number of periods but also with the coupling constraints (Alsaç et al. 2004).

2.4.5.4 Time-Constrained Economic Dispatch

Classical economic dispatch problems are static in the sense that only one snapshot problem is solved at a time without taking into consideration the dynamic nature of system conditions, such as instantaneous changes in system demand or changes in the network topology. In real-time operation, some controls in the grid can react very quickly to this kind of dynamic situation while other controls may not be able to respond to the problem as quickly and adequately as possible.

For a real-time operation, control decisions should take into consideration system conditions for the next hour or eventually for the next two hours. For short-term planning, control decisions should take into consideration the system conditions for the next 24 h or even the next week. If there is no constraint between these consecutive hours, then the economic dispatch problems can be solved independently. However, nowadays it becomes essential to include temporal restrictions, such as generation units' ramp constraints, and objective variables that cannot move more than a specified value or have a minimum movement among adjacent periods. There are two approaches to solve this type of problem: (1) solve all periods together as a large optimization problem or (2) apply decomposition methods to include time-coupling constraints (David & Li 1993).

2.5 Conclusions

With the wide-spread adoption of the Internet throughout the world, the inundation of data, lots of them, is inevitable. Almost every business, institution, and organization is and will be affected by this growing wave of big data. Those organizations have no choice but to be ready to deal with that phenomenon. Health care, finance, social networking, oil and gas, energy, and even power systems are some major areas which will face this sea change. In the first part of this chapter, the general problem of big data is stated, in the context of some of these business areas. Then, the readers are introduced some of the latest methods and algorithms used in processing and analyzing such big data. Those algorithms come from machine learning, such as ANNs, SVM, decision-tree algorithms. Models from deep learning, such as CNNs, RNN, and DBN, are also introduced. The final part of this chapter describes some of the development and challenges facing both the traditional power systems and new SG environment. Some of the developments include the growing adoption of PMU, and associated challenges of processing large amount of data produced by those devices. Optimization problems associated with big data will also be more complicated and challenging going forward. To solve such problems, decomposition methods have become popular as they are effective in solving varied and complex problems. The chapter concludes with outlining some emerging problems in new power systems.

References

Aggarwal, S.K., Saini L.M. & Kumar A. (2009). Electricity price forecasting in deregulated markets: A review and evaluation. *International Journal of Electrical Power & Energy Systems*, vol. 31, no. 1, pp. 13–22.

Alpaydin, E. (2014 July–August). *Introduction to Machine Learning.* Cambridge, MA: The MIT Press.

Alsaç, O., Bright, J.M., Brignone, S., Prais, M., Silva, C., Stott, B. & Vempati, N. (2004). The rights to fight price volatility. *IEEE Power and Energy Magazine*, vol. 2, no. 4, pp. 47–57.

Bengio, Y. (2009). Learning Deep Architectures for AI. *Foundation and Trends. R. O in Machine Learning*, vol. 2, no. 1, pp. 1–127.

Bengio, Y. (2013). Deep learning of representations: Looking forward. *1st International Conference on Statistical Language and Speech Processing. SLSP'13*, Tarragona, Spain, pp. 1–37.

Bihl, T.J., Young II, W.A. & Weckman, G.R. (2016). Defining, understanding, and addressing big data. *International Journal of Business Analytics (IJBAN)*, vol. 3, no. 2, pp.1–32.

David, A.K. & Li, Y.Z. (1993 February). Effect of inter-temporal factors on the real time pricing of electricity. *IEEE Transactions on Power Systems*, vol. 8, no. 1, pp. 44–52.

Deepak, S. & Swarup, K.S. (2011). Electricity price forecasting using artificial neural networks. *International Journal of Electrical Power & Energy Systems*, vol. 33, no. 3, pp. 550–555, March 2011.

Foruzan, E., Sangrody, H., Lin, J. & Sharma, D.D. (2017 September). Fast sliding detrended fluctuation analysis for online frequency-event detection in modern power systems. *North American Power Symposium (NAPS)*, West Virginia, pp. 1–6.

Gangammanavar, H., Sen, S. & Zavala, V.M. (2016 March). Stochastic optimization of sub-hourly economic dispatch with wind energy. *IEEE Transactions on Power Systems*, vol. 31, no. 2, pp. 949–959.

Hasim, S., Senior, A.W. & Beaufays, F. (2014). Long short-term memory recurrent neural network architectures for large scale acoustic modelling. *Fifteenth Annual Conference of the International Speech Communication Association*, Interspeech, 2014.

He, X., Ai, Q., Qiu, R., Huang, W., Piao, L. & Liu, H. (2017 March). A big data architecture design for smart grids based on random matrix theory. *IEEE Transactions on Smart Grid*, vol. 8, no. 2, pp. 674–686.

Hijazi, S., Kumar, R. & Rowen, C. (2015). Using convolutional neural networks for image recognition. Cadence. https://ip.cadence.com/uploads/901/cnn_wp-pdf.

Hinton, G. (2002). Training products of experts by minimizing contrastive divergence. *Neural Computing*, vol. 14, no. 8, pp. 1771–1800.

Hinton, G., Deng, L., Yu, D., Mohamed, A-R, Jaitly, N, Senior, A., Vanhoucke, V., Nguyen, P., Sainath, T., Dahl, G. & Kingsbury, B. (2012). Deep neural networks for acoustic modeling in speech recognition: The shared views of four research groups. *IEEE Signal Process Magazine*, vol 29, no. 6, pp. 82–97.

Hinton, G. & Salakhutdinov, R. (2006). Reducing the dimensionality of data with neural networks. *Science*, vol. 313, no. 5786, pp. 504–507.

Hong, T., Chen, C., Huang, J., Lu, N., Xie, L. & Zareipour, H. (2016 September). Big data analytics for grid modernization. *IEEE Transactions on Smart Grid*, vol. 7, no. 5, pp. 2395–2396.

Hu, J. & Vasilakos, A. (2016 September). Energy big data analytics and security: Challenges and opportunities. *IEEE Transactions on Smart Grid*, vol. 7, no. 5, pp. 2423–2436.

Kalogirou, S.A. (2000 September). Applications of artificial neural-networks for energy systems. *Applied Energy*, vol. 67, no. 1–2, pp. 17–35.

Krizhevsky, A., Sutskever, I. & Hinton, G.E. (2009). ImageNet classification with deep convolutional neural networks. *Advances in Neural Information Processing Systems*, vol. 22, pp. 1097–1105.

Krizhevsky, A., Sutskever, I. & Hinton, G.E. (2012). ImageNet classification with deep convolutional neural networks. *Advances in Neural Information Processing Systems*, vol. 25, pp. 1106–1114.

Magnago, F., Zhang, L. & Nagarkar, R. (2015 September). Three phase distribution state estimation utilizing common information model. *2015 IEEE Eindhoven PowerTech*, Eindhoven, Netherlands, pp. 1–6. doi: 10.1109/PTC.2015.7232515.

Negnevitsky, M., Mandal, P. & Srivastava, A.K. (2009). Machine learning applications for load, price and wind power prediction in power systems. *15th International Conference on Intelligent System Applications to Power Systems*, 8–12 Nov. 2009, Curitiba, Brazil, pp. 1–6.

Pinto, H., Magnago, F., Brignone, S., Alsac, O. & Stott, B. (2006). Security constrained unit commitment: Network modeling and solution issues. *IEEE PES Power Systems Conference and Exposition*, Atlanta, GA, pp. 1759–1766.

Raoufat, M.E., Tomsovic, K. & Djouadi, S.M. (2016 November). Virtual actuators for wide-area damping control of power systems. *IEEE Transactions on Power Systems*, vol. 31, no. 6, pp. 4703–4711.

Raoufat, M.E., Tomsovic, K. & Djouadi, S.M. (2017 November). Dynamic control allocation for damping of damping inter-area oscillations. *IEEE Transactions on Power Systems*, vol. 32, no. 6, pp. 4894–4903.

Roman, T. (2004). Classification and regression trees (CART) theory and applications. Diss. Humboldt University, Berlin.

Rutkowski, L., Jaworski, M., Pietruczuk, L. & Duda, P. (2014). The CART decision tree for mining data streams. *Information Sciences*, vol. 266, pp. 1–15.

Saini, L.M., Aggarwal, S.K. & Kumar, A. (2010). Parameter optimization using genetic algorithm for support vector machine-based price-forecasting model in national electricity market. *Generation, Transmission & Distribution, IET,* vol. 4, no. 1, pp. 36–49.

Scherer, D., Müller, A. & Behnke, S. (2010). Evaluation of pooling operations in convolutional architectures for object recognition. *Proceedings of International Conference on Artificial Neural Networks (ICANN),* 15–18 Sep. 2010, Thessaloniki, Greece, pp. 92–101.

Soares, T., Fernandes, F., Morais, H., Faria, P. & Vale, Z. (2012 May). ANN-based LMP forecasting in a distribution network with large penetration of DG. *IEEE, PES Transmission and Distribution Conference and Exposition (T&D),* pp. 1–8.

Srivastava, N. & Salakhutdinov, R. (2012). Multimodal learning with deep Boltzmann machines. *Advances in neural information processing systems (NIPS),* 03–08 Dec. 2012, Harrahs and Harveys, Lake Tahoe.

Wang, C. & Shahidehpour, S.M. (1992 November). A decomposition approach to non-linear multi-area generation scheduling with tie-line constraints using expert systems. *IEEE Transactions on Power Systems,* vol. 7, no. 4, pp. 1409–1418.

Wu, X., Zhu, X., Wu, G. & Ding, W. (2014). Data mining with big data. *IEEE Transactions on Knowledge and Data Engineering,* vol. 26, no. 1, pp. 97–107.

Xue-Wen, C. & Lin, X. (2014). Big data deep learning: Challenges and perspectives. *IEEE Access,* vol. 2, pp. 514–525.

Zhi-Hua, Z., Chawla, G.J. & Yaochu, J.W. (2014). Big data opportunities and challenges: Discussions from data analytics perspectives [discussion forum]. *IEEE Computational Intelligence Magazine,* vol. 9, no. 4, pp. 62–74.

3

The Role of Big Data in Smart Grid Communications

Francisco M. Portelinha Júnior
National Institute of Telecommunications

Denisson Q. Oliveira
Federal University of Maranhão

CONTENTS

3.1   Introduction...37
3.2   The Grid Modernization...38
3.3   The Grid Interconnection with the Internet of Things....................39
3.4   Data Traffic Pattern in a Smart Grid Environment..........................42
     3.4.1   Phasor Measurement Unites Applied to Distribution Systems....43
     3.4.2   Advanced Metering Infrastructure (AMI).............................44
3.5   The Massive Flow of Information in a Smart Scenario.....................45
3.6   The Volume of Generated Data in a Smart Distribution System:
     A Case of Study...47
     3.6.1   The Simulated Case I—Generated Data by PMUs....................48
     3.6.2   Case II—Generated Data by Metering Infrastructure 48
3.7   Conclusion 50
References..51

3.1 Introduction

A smart grid can be described as a vast sensor network, with a large variety of connected devices. The growth in the number of smart devices and the increase in operational requirements will raise the flow of information inside the network. Therefore, how to quantify and analyze these data arises as one big concern to enhance grid operation. However, Big Data rises as a proper technique to efficiently manage and take profit from this large volume of data. This promising tool will give operators and utilities a better understanding of customer behavior, demand consumption, weather forecast, power outages, and failures. Also, the deployment of robust methodologies will help to turn the grid smarter. However, it is vital to quantify the volume of produced data by these devices and how to take advantage of them. Therefore, this chapter aims to characterize and to evaluate the emerging growth of data in communications network applied to smart grid scenarios. A future active distribution system will serve as an example to demonstrate the massive volume of generated data by intelligent devices to control and monitor the grid.

3.2 The Grid Modernization

The electric power system behaves dynamically. Any variation in the power generation will affect the state of customers' power supply. As a result, there is significant concern about how the system will respond to these abrupt variations and how it can be monitored and operated in real time (Amin and Wollenberg, 2005).

The concept of Smart Grid deploys the integration of Information and Communication Technologies (ICTs) together with the electrical system (Li et al., 2014). As a way to accomplish the pursued smartness, this future power system will make use of ICTs to increase the reliability, robustness, safety, reduction of losses, and failures within the electrical system, mainly in distribution level, where 90% of the failures occur (Farhangi, 2010; Fang et al., 2012).

Moreover, the growth of the penetration of distributed generation, mainly of solar and wind energy sources, requires the increase in the monitoring capacity of the electrical network (Bouhafs et al., 2012). Thus, the efficient exchange of information between all participating agents becomes fundamental and must be guaranteed.

The automation of the electric power system for real-time processing relies on a different amount of intelligent electronic devices (IEDs), which are responsible for capturing data from several sources as (Jiang et al., 2016):

- User's profile history;
- Phasor measurement unit (PMU) measurements;
- Distributed energy resources;
- Advanced metering infrastructure (AMI);
- Sensors;
- Actuators;
- Breakers;
- Capacitor banks;
- Utilities.

As a way to incorporate all these new technological advances, AMI is pointed out as the first step in grid modernization (Bouhafs et al., 2012). It will provide a two-way communication system between utilities and the users (industrial/commercial/residential). It will also be responsible not only for metering issues but also for demand response (DR), supporting state estimation, pricing issues, and taking advantages of real-time communication to deliver data instantaneously to manage balance and supply of the system (Sun et al., 2016).

Also, in future distribution systems, PMU will be applied to control applications and functions including system protection, state estimation, voltage/frequency control, islanding monitoring, renewable resources control, and islanding operation (Sánchez-Ayala et al., 2013; Liu et al., 2012).

All these devices will generate and exchange a massive amount of information to achieve grid autonomously. By deploying a massive variety of intelligent sensors to monitor and control grid operation status, the volume of produced information will rise. Therefore, to achieve better performance on system operation, it is essential to properly manage and exploit the data sets from those resources (Daki et al., 2017; Asad et al., 2017), which is not a trivial task. As a way to make a profit from these challenges, new computational technologies and data management analytical tools, such as Big Data, can be a solution (Daki et al., 2017).

The use of Big Data will help utilities and operators to increase their profitability by accurately managing their massive amount of generated data. As a first step to understand all the massive flow of information inside an automated system, it is crucial to quantify and identify accurately which devices are going to generate data and how much data are going to be generated by each device at a specific task to make use of Big Data technologies to improve grid robustness.

3.3 The Grid Interconnection with the Internet of Things

Integrated, with high-performance, highly reliable, robust and flexible are the characteristics of an intelligent communication network. It will be responsible for data collection, routing, monitoring, and management of all active devices in the network (Fadlullah et al., 2011).

As an example, to understand the communication requirements between smart devices, consider an integrated power network with a deployed advanced communication network, where thousands of devices will be sending messages to hundreds of substations, which will be connected to a variety of control centers responsible for decision-making. The size of the network will be enormous and with almost no human interaction (Wang and Fapojuwo, 2017). Each distribution system will have smart meters, PMUs, and other IEDs ranging from a few hundred to a few thousand, according to the extent of their geographic area.

In this scenario, the Internet of Things (IoT) concept becomes an essential ally in the constant development of intelligent networks (Gazis, 2017). Applying the IoT concept within electrical power systems is a trend for the flexibility and scalability between all players inside the grid. As a way, to better illustrate this interconnection, Figure 3.1 illustrates the idea of a cyber–physical system, which is a highly automated distribution system connected to an advanced communication network (Yu and Xue, 2016).

The advanced communication network as shown in Figure 3.1 will provide the link between controlling the grid and system autonomy. The robust and flexible communication infrastructure is part of the system integration, which comprises some critical requirements for system operation, as listed below (Gungor et al., 2011, 2013):

- Latency: By definition the network delay or the expression of how long it takes for a packet of data to travel from one network point to another;
- Bandwidth: The concept of bandwidth is fundamental in determining the communication requirements for intelligent communication networks since it is a factor that directly influences the choice of technology (for example, wired or wireless communication);
- Interoperability/flexibility: It is defined as the ability of multiple systems to work together and be compatible with each other;
- Data throughput: It is the ability to transfer information at the maximum data rate;
- Cyber security: Different protocol standards will flow across the network, authentication, authorization, and privacy requirements are critical issues.

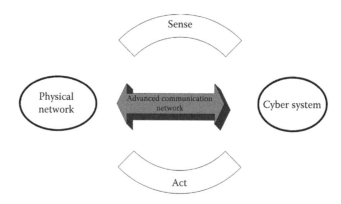

FIGURE 3.1
Smart grid as a cyber–physical system. Adapted from Yu and Xue (2016).

With this high dependence on information services, the demand for new technologies increases (Portelinha et al., 2016). As these devices need to communicate with each other, the dependence on the integration of smart networks and communication networks grows, so the grid must adapt to these new integration challenges (Portelinha et al., 2017).

All of these devices produce/exchange different kinds of data, such as environmental data, geographical data, operation data, weather data, and customer data. The massive amount of data generated by these IoT devices can be characterized as a big data set. This big data set is going to be described according to their heterogeneity, variety, unstructured feature, noise, and high redundancy (Marjani et al., 2017).

The interconnection between IoT devices, within an advanced communication network, and the connection with Big Data technologies is depicted in Figure 3.2 (Marjani et al., 2017). Many technologies can fulfill all these strict requirements of the communication link between the electrical grids (Gungor et al., 2011). The choice of a communication technology should be based on the need for reliability, security, and availability of each service offered. The type of operation to be performed is another issue that should be taken into account. Those considered critical, such as control and critical operation, require a more robust network infrastructure.

The communication infrastructure must adapt its actual configuration with low investment to fulfill future needs. Otherwise, the stakeholders can consider it unfeasible. Thus, existing technologies must be considered as solutions, decreasing the implementation costs and enabling the application in smart grids.

Some technologies are presented in Gungor et al. (2011, 2013; Aravinthan et al., 2011; Ho et al., 2013). However, there are advantages and limitations of a variety of prominent technologies. Wired communications offer high data

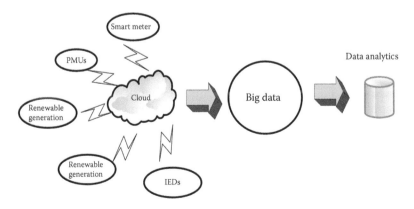

FIGURE 3.2
The integration of IoT and Big Data technology. Adapted from Marjani et al. (2017).

rates, but with low mobility and high infrastructure cost. Mobile technologies are prominent to be applied in smart grid scenarios. They are robust, flexible, scalable, and more important, new standards have been proposed to support machine-type communication, offering more power efficiency, lower operation costs, and flexibility (Fapujwo, 2017). These new standards emerge as a solution to make the use of smart devices feasible within wideband mobile communication in the licensed spectrum.

Those standards are mainly extended from existing LTE (Long Term Evolution) functionalities and must satisfy strict M2M (machine-to-machine) requirements. NB-IoT (narrow band IoT) networks are part of the proposed release 13 from the Third Generation Partnership Project (3GPP) (TS 136401, 2010). In this version, several features have been introduced, focusing on the machine communication functionalities, such as efficient spectrum utilization, coverage improvement, low-cost devices and high capacity, but better described in version 14, focused on enhancing coverage (Ratasuk et al., 2016).

3.4 Data Traffic Pattern in a Smart Grid Environment

The actual electrical system is built as a unidirectional system, with a little intelligence and without the capacity to transmit information in real time (Bouhafs et al., 2012). Advanced communication systems are essential for protection, control, and monitoring of the grid. The current power grid monitoring is based on systems such as SCADA (Supervisory Control and Data Acquisition) and AMR (Automatic Meter Reading), which are not suitable for future needs of a smart grid scenario (Lai and Lai, 2015). As a result, advanced measurement systems must be implemented with the objective of providing real-time communication in a bidirectional way, providing increased robustness, reliability, and security.

Future distribution systems should be provided with reliable and flexible network infrastructure, with strict requirements. The primary functions of this system are to monitor and sense (Marjani et al., 2017). Both are related to the perception of any change and acting/reacting to this change, as described in Figure 3.2. The distribution network must be provided with strict applications requirements to improve grid smartness, as described as follows (Fan et al., 2012):

- Distribution Control and Protection: Critical communications are the primary functions. IED devices are responsible for locating and detecting faults, and exchanging reporting messages;
- Wide Area Monitoring System: The system will collect information from large areas and substations and make critical decisions;

- DR: Several sources of distributed energy resources will be connected to the system, which brings more intermittent variables to be controlled and monitored;
- AMI: Smart meters will play an essential role in this future grid. Besides billing, they will be responsible for consumer interaction, load control, DR, islanding detecting and other functionalities.

Requirements in the physical layer will diverge on data transmission, latency, and user's priority. Some critical communication requirements for each listed application regarding latency and frame size (Kuzlu et al., 2014) are better illustrated in Table 3.1.

Future power networks must be integrated with an AMI to support smart applications. The grid must support other functionalities, besides metering, such as the possibility to choose from whom or when to buy energy from a given utility, choose if it is time to use their private energy resources, especially during peak hours, to monitor load demands, and billing (Sánchez-Ayala et al., 2013).

This vast smart distribution network will generate a large quantity of data. For instance, PMUs used different sample rates and AMI system might collect data every 1–15 min or even hours (Daki et al., 2017). In future active distribution networks, AMI and PMUs will work together to keep grid reliability and robustness. Therefore, it is crucial to measure and analyze the amount of generated data by this smart system to better understand the grid performance and how big data techniques will transform the grid modernization.

3.4.1 Phasor Measurement Unites Applied to Distribution Systems

PMUs are going to be placed at strategic locations and will perform precise voltage and current phasor measurements due to their interconnection with the global positioning system (GPS) (Liu et al., 2012). The correct acquisition of voltage and current measurements allows the operator to estimate the state of the electrical system accurately.

The measurements performed by the PMUs can be defined in measurements per cycle (10, 20, 30, and 60 are the most used). These data are sent to concentrators named Phasor Data Unit (PDU), where the data will be treated

TABLE 3.1

Smart Applications Requirements

Application	Latency	Message Size (Bytes)
Protection	1–10 ms	Few
Control	100 ms	Few
Monitoring	1 s	Few–Medium
Metering	Min–Hours	Medium

for dynamic issues. The latency and frequency of measurement of each application will significantly influence the necessary bandwidth to transmit and receive data. Therefore, it is essential to determine the minimum bandwidth to support PMUs communication data, which can be easily calculated by

$$BW = N_{frame} \times f_s \times N_{PMU} \tag{3.1}$$

where N_{frame} is the frame size in Bytes, f_s is the sampling frequency, and N_{PMU} is the number of connected PMUs.

The sampling frequency of each application will significantly influence the total transmission capacity of the communication network. Because of these new grid requirements, the volume of synchrophasors installed in the distribution system will grow, and the amount of generated data to be analyzed will grow.

3.4.2 Advanced Metering Infrastructure (AMI)

Through smart meters deployment, several smart applications will be possible always in association with the appropriate communication infrastructure. The number of smart meter's action is vast, and one primary concern is how to take advantage of this enormous amount of data. Each kind of measurement of the smart meter has different size and sample rate. As a way to estimate the traffic volume, it is essential to know each kind of message that will be sent and how many smart meters the AMI's infrastructure must support. In Luan et al. (2010), it is shown the traffic messages profile of an automated infrastructure, the size, and sample rate of each transmitted message from a smart meter.

Taking it into account, it is possible to evaluate the volume of generated data by each smart meter. This measurement device will be placed in every house, commercial building, and industrial facility; the number of devices is going to be huge, such as the flow of information to control and diagnosis grid operation.

Therefore, one critical issue is to estimate the accurate number of smart meter inside the distribution infrastructure. However, to evaluate this factor, the communication network design must be considered. Some issues related to propagation losses, receiver and transmitter antenna gain, and geographical issues must be calculated. The most important feature is the distance coverage factor, which is influenced by the geographical size where one radio base station can provide connectivity. Keeping this in mind, the estimated number of smart meters can be calculated as follows (Persia et al., 2015):

$$N_{SM} = \rho \pi d^2 \tag{3.2}$$

where ρ is the smart meter density (number of smart meters per square meter), which relies on the typical geographical scenario (urban, suburban, and rural). In critical operation, a minimum data rate of 64 kbps per meter is required.

3.5 The Massive Flow of Information in a Smart Scenario

The ongoing growth of IEDs to achieve grid smartness will generate a massive amount of data from those connected devices. Researchers have shown that the electric utilities had generated already some hundreds of millions of gigabytes of data in their systems, and it is increasing to terabytes of collected data. Only syncrophasors alone have produced hundreds of terabytes of data per year (Asad et al., 2017).

The data generated from these devices present an immense opportunity. Analyzing this massive set of data will help to increase grid reliability, and enable applications features, such as predictive analytics, demand-side management, real-time grid awareness, outage detection, asset management, and theft detection (Asad et al., 2017). Figure 3.3 illustrates the role of big data in a smart grid scenario.

The use of big data will enhance the usability of the generated data to make a better prediction, management, and processing (Jiang et al., 2016). Several fields in a smart grid scenario can improve their operation by recognizing data patterns. Many applications can take profit of big data. The most important features for a future active distribution network are listed below (Jiang et al., 2016):

- DR: Predicting and analyzing the user's patterns will help to predict power demand accurately;
- Distributed Energy Resources: Forecasting and accurate schedule load are essential to energy planning. New intermittent sources of energy will be integrated into grid extending the complexity of grid operation;

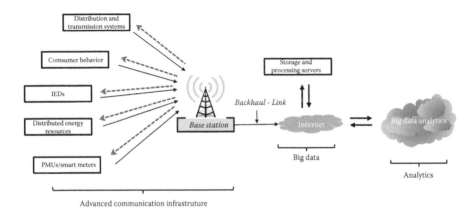

FIGURE 3.3
The integration of information and communication technologies in a smart grid context.

- AMI: Exploring the generated data from smart meters will help utilities to identify customer's patterns, load forecast, energy demand, and demand-side management;
- Distribution Automation: Sensing and monitoring the power distribution system will help to increase grid robustness by predicting outage situations.

The foundation of data technology comprises five constraints: volume, velocity, variety, value, and veracity (Subhani et al., 2015) as depicted in Figure 3.4.

Advanced data analytics is done with mathematical techniques including predictive analytics, data mining, artificial intelligence, and fuzzy theory (Marjani et al., 2017). The application of these technologies enables optimal decision-making by exploring big data sets (Jiang et al., 2016). The examination process of these sets will transform this considerable volume of data into a more readable data and metadata format for analytical procedures (Subhani et al., 2015). Understanding these data will help utilities and stakeholders to make efficient decisions and turn to a more profitable grid.

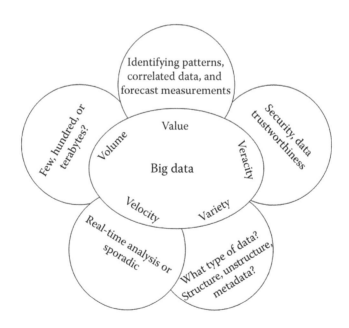

FIGURE 3.4
The 5 V's of big data in a smart grid scenario. Adapted from Subhani et al. (2015).

3.6 The Volume of Generated Data in a Smart Distribution System: A Case of Study

This section aims to analyze the amount of information generated by the devices responsible for controlling, monitoring, and management of a future automated distribution system.

In normal conditions, smart applications operate within a traffic schedule. For example, metering messages are sent every 15 min, none aperiodic control and outage message is sent if it is not necessary. Therefore, to measure the total amount of data, the critical operation environment must be considered. In harsh environments, the grid will continuously send messages, perform load flow, demand response, to name a few operations.

For simulation purposes, a modified version of the IEEE 123 bus system is considered, as shown in Figure 3.5. Here, the PMU placement step has already been done as in Jamil et al. (2014).

The IEEE 123 distribution model has been chosen because of its high load topology, which must comprise a considerable number of intelligent devices. For all simulations, several IEDs, PMUs, and smart meters will capture data to and send to the MGCC, where decisions will be made.

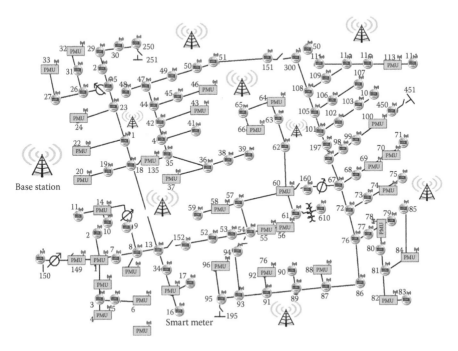

FIGURE 3.5
Modified IEEE 123 distribution systems.

The communication infrastructure is based on wireless technology. Cellular technologies are promising for this type of application due to its low infrastructure cost, high coverage distance, and support for machine-type communications. The NB-IoT standard has been adopted to support communication in this highly automated system. The network dimensioning and implementation are not the focus of this section and are better described in Portelinha et al. (2018).

3.6.1 The Simulated Case I—Generated Data by PMUs

The volume of data generated by synchronous phasor throughout the system will depend on the number of PMUs installed, the size of the message sent, and the sampling frequency at which the data are captured.

In this case study, 49 synchrophasors will be placed according to Jamil et al. (2014). This number is the optimal number for the 123-bus system node. Each PMU sends a frame packet at a fixed sampling rate, according to IEEE Std. C37.118 (IEEE, 2011); the size of the data message forwarded per packet is 80 fixed bytes, plus the fields corresponding to each phasor, transducer, and the digital signal for the formation of the message packet.

An important factor is the sampling rate, which is dependent on the frequency of the distribution system. In the case of the Brazilian electrical system, 60 Hz is considered. The acquisition rate of these samples is defined according to the digital-analog converter at the input of the PMU. The sampling rates most commonly used for capture are 10, 20, 30, and 60 synchrophasors per second and will be used in this work as the basis for the calculation of the minimum bandwidth requirements. In this way, it is possible to amount of generated data by each synchrophasor, by Equation (3.1) and considering that each PMU is comprised of eight phasor channels and two digital channels for different types of synchrophasor sampling rates.

Table 3.2 shows the volume of information generated by phasor measurements at different sampling rates. For example, for the case study of Figure 3.5, to accommodate 49 PMUs, at a sampling rate of 10 measures per second, the total data generated are around some hundreds of megabits per second. As can be seen from Table 3.2, the volume of generated data from synchrophasors is exceptionally high, and it gets higher if more samples are used to quantize the data. Due to the high amount of data generated by the PMUs, the use of big data becomes clear. This massive volume of generated data for optimizing dynamic tasks must be organized, and better treated, as a way to achieve better performance and grid reliability.

3.6.2 Case II—Generated Data by Metering Infrastructure

It is necessary to deploy a variety of sensors along the distribution grid to accomplish all the required smartness. These sensors will constitute the AMI. The collected data must be sent to the distribution system operator

TABLE 3.2

Volume of Generated Data by PMUs

#PMU	#Bytes	Sampling Rate (Mbps)					
		10	12	15	20	30	60
1	112	8.96	10.752	13.44	17.92	26.88	53.76
2	224	16.06	19.27	24.08	32.11	48.17	96.34
3	336	43.16	51.79	64.74	86.32	129.48	258.96
6	672	232.02	278.43	348.04	464.05	696.07	1392.15
9	1008	1871.05	2245.26	2806.57	3742.10	5613.15	11226.29
10	1120	16764.60	20117.51	25146.89	33529.19	50293.79	100587.57
15	1680	225316.16	270379.39	337974.24	450632.32	675948.47	1351896.95
20	2240	4037665.548	4845198.66	6056498.323	8075331.097	12112996.6	24225993.29
30	3360	108532449.9	130238940	162798674.9	217064899.9	325597350	651194699.7
49	5488	4765008682	5718010419	7147513023	9530017364	1,4295E+10	28590052093
50	5600	2,13472E+11	2,5617E+11	3,20209E+11	4,26945E+11	6,4042E+11	1,28083E+12
60	6720	1,14763E+13	1,3772E+13	1,72144E+13	2,29526E+13	3,4429E+13	6,88577E+13

center to perform the necessary actions and send feedback signals to the corresponding actuators.

In Figure 3.5, smart meters are assumed to be randomly placed within the whole area, and the maximum allowable number of metering devices is obtained from Equation (3.2). The optimum allocation of smart meters is not the focus of this research.

For a fair comparison, the maximum data rate of 64 kbps per meter is assumed (Persia et al., 2015), such as in critical mission scenario. As a way to determine the minimum number of smart meters to the desired distribution system, it is essential to take into account parameters such as the geographical size, the chosen technology, the distance between the transmitter and receiver, antennas gain, propagation losses, etc., as in Portelinha et al. (2018). To find a systematic model, the authors suggest to the reader to go through the references (Portelinha et al., 2016, 2017). In this simulated scenario, the focus is to determine the number of smart meters to be deployed in Figure 3.5. Therefore, the coverage factor of one base station must be evaluated. Moreover, this simulation aims to determine the coverage area of one base station and evaluate the minimum and the maximum number of supported devices within the range of one radio station.

Given the simulated scenario in Figure 3.5, assume an urban scenario pattern. Figure 3.6 shows the relationship between the connectivity area of one base station and the number of possible devices within this determined area.

In Figure 3.6, it is notable the rise of smart meters devices, as the coverage factor increases. For example, if we assume a 0.5 km² coverage ratio, one base station can provide connectivity to almost 800 smart meters. To provide connectivity to the whole system illustrated in Figure 3.5, at least five base

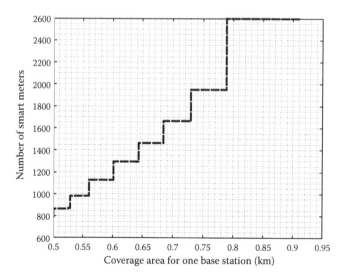

FIGURE 3.6
The influence of the distance factor to determine the number of smart meters.

stations will be needed, and around 4000 smart meters will be supported since this scenario has a geographical area of approximately 2 km².

Furthermore, as the number of deployed smart meters rises, the flow of information inside the distribution also increases exponentially. However, all this flow of information must be appropriately treated by advanced analytics methods.

Therefore, it is easy to understand that one critical issue in future power system design is how to take advantage of all these generated data and efficiently operate both the electrical and communication flow. More interconnected devices mean more support to better estimate and operate the grid. However, this enormous amount of generated information can be prejudice if not treated correctly.

3.7 Conclusion

In the operation of future power networks, it is essential to understand the flow of data inside the grid. These data are useful information when combined with advanced analytic technologies to predict the electrical system operation status.

Despite that, grid modernization comes with a considerable amount of IEDs that will be deployed inside the grid with operational requirements. The data generated from these devices, such as by smart meters and PMUs, result in a set of Big Data. The expansion of the intelligence of the grid will exponentially increase the Big data sets.

In this work, we have described a few problems to enable a smart grid environment in a big data scenario. Utilities must learn how to handle and transform this massive volume of generated data by thousands of devices to enhance grid reliability and robustness.

References

Amin, S. M. and Wollenberg, B. F. Toward a smart grid: Power delivery for the 21st century. *IEEE Power and Energy Magazine* 3 no. 5 (2005): 34–41.

Aravinthan, V.; Karimi, B.; Namboodiri, V. and Jewell, W. Wireless communication for smart grid applications at distribution level—Feasibility and requirements. *IEEE Power and Energy Society General Meeting*, San Diego, CA, 2011.

Asad, Z. and Rehman Chaudhry, M. A. A two-way street: Green big data processing for a greener smart grid. *IEEE Systems Journal* 11 no. 2 (2017): 784–795.

Bouhafs, F.; Mackay, M. and Merabti, M. Links to the future: Communication requirements and challenges in the smart grid. *IEEE Power and Energy Magazine* 10 no. 1 (2012): 24–32.

Daki, H.; El Hannani, A.; Aqqal, A. J.; Haidine, A. and Dahbi, A. Big Data management in smart grid: concepts, requirements and implementation. *Journal of Big Data* 4 no. 1 (2017): 13.

Fadlullah, Z. M.; Fouda, M. M.; Kato, N.; Takeuchi, A.; Iwasaki, N. and Nozaki, Y. Toward intelligent machine-to-machine communications in smart grid. *IEEE Communications Magazine* 49 no. 4 (2011): 60–65.

Fan, Z.; Chen, Q.; Kalogridis, G.; Tan, S. and Kaleshi, D. The power of data: Data analytics for M2M and smart grid. *3rd IEEE PES Innovative Smart Grid Technologies Europe (ISGT Europe)*, Berlin, 2012.

Fang, X.; Misra, S.; Xue, G. and Yang, D. Smart grid—The new and improved power grid: A survey. *IEEE Communications Surveys & Tutorials* 14 no. 4 (2012): 944–980.

Farhangi, H. The path of the smart grid. *IEEE Power and Energy Magazine* 8 no. 1 (2010): 18–28.

Gazis, V. A survey of standards for machine-to-machine and the internet of things. *IEEE Communications Surveys & Tutorials* 19 no. 1 (2017): 482–511.

Gungor, V. C. et al. Smart grid technologies: Communication technologies and standards. *IEEE Transactions on Industrial Informatics* 7 no. 4 (2011): 529–539.

Gungor, V. C. et al. A survey on smart grid potential applications and communication requirements. *IEEE Transactions on Industrial Informatics* 9 no. 1 (2013): 28–42.

Ho, Q. D.; Gao, Y. and Le-Ngoc, T. Challenges and research opportunities in wireless communication networks for the smart grid. *IEEE Wireless Communications* 20 no. 3 (2013): 89–95.

IEEE Standard for Synchrophasor Measurements for Power Systems, IEEE Std C37.118.1–2011 (Revision of IEEE Std C37.118–2005) (2011): 1–61.

Jamil, E.; Rihan, M. and Anees, M. A. Towards optimal placement of phasor measurement units for smart distribution systems. *6th IEEE Power India International Conference (PIICON)*, Delhi, 2014.

Jiang, H.; Wang, K.; Wang, Y.; Gao, M. and Zhang, Y. Energy big data: A survey. *IEEE Access* 4 (2016): 3844–3861.

Kuzlu, M.; Pipattanasomporn, M. and Rahman, S. Communication network requirements for major smart grid applications in HAN, NAN, and WAN. *Computer Networks* 67 (2014): 74–88.

Lai, C. S. and L. L. Lai. Application of Big Data in Smart Grid. *IEEE International Conference on Systems, Man, and Cybernetics*, Kowloon, 2015.

Li, H.; Dimitrovski, A.; Song, J. B.; Han, Z. and Qian, L. Communication infrastructure design in cyber-physical systems with applications in smart grids: A hybrid system framework. *IEEE Communications Surveys & Tutorials* 16 no. 3 (2014): 1689–1708.

Liu, J.; Tang, J.; Ponci, F.; Monti, A.; Muscas, C. and Pegoraro, P. A. Trade-Offs in PMU deployment for state estimation in active distribution grids. *IEEE Transactions on Smart Grid* 3 no. 2 (2012): 915–924.

Luan, W.; Sharp, D. and Lancashire, S. Smart grid communication network capacity planning for power utilities. *IEEE PES T&D*, New Orleans, LA, 2010.

Marjani, M. et al. Big IoT data analytics: Architecture, opportunities, and open research challenges. *IEEE Access* 5 (2017): 5247–5261.

Persia, S.; Petrini, V.; Rea, L. and Valenti, A. Wireless M2M capacity analysis for smart distribution grids. *AEIT International Annual Conference (AEIT)*, Naples, 2015.

Portelinha Júnior, F. M.; de Souza, A. C. Z.; Castilla, M.; Queiroz Oliveira, D. and Ribeiro, P. F. Control strategies for improving energy efficiency and reliability in autonomous microgrids with communication constraints. *Energies* 10 (2017): 1443.

Portelinha, F.; Oliveira, D. Q.; de Souza, A. C. Z.; Ribeiro, P. F.; de Nadai, B. and Marujo, D. The impact of electric energy consumption from telecommunications systems on isolated microgrids. *5th IET International Conference on Renewable Power Generation (RPG)*, London, 2016.

Portelinha Júnior, F. M.; Zambroni de Souza, A. C.; Ribeiro, P. F.; Oliveira, D. Q. and de Nadai Nascimento, B. Design and performance of an advanced communication network for future active distribution systems. *Journal of Energy Engineering* 144 (2018): 04018019.

Ratasuk, R.; Vejlgaard, B.; Mangalvedhe, N. and Ghosh, A. NB-IoT system for M2M communication. *IEEE Wireless Communications and Networking Conference Workshops (WCNCW)*, Doha, 2016.

Sánchez-Ayala, G.; Agüerc, J. R.; Elizondo, D. and Lelic, M. Current trends on applications of PMUs in distribution systems. *IEEE PES Innovative Smart Grid Technologies Conference (ISGT)*, Washington, DC, 2013.

Subhani, S.; Gibescu, M. and Kling, W. L. Autonomous control of distributed energy resources via wireless machine-to-machine communication: A survey of big data challenges. *IEEE 15th International Conference on Environment and Electrical Engineering (EEEIC)*, Rome, 2015.

Sun, Q. et al. A comprehensive review of smart energy meters in intelligent energy networks. *IEEE Internet of Things Journal* 3 no. 4 (2016): 464–479.

TS 136 401- V9.2.0- LTE; Evolved Universal Terrestrial Radio, Access Network (E-UTRAN); Architecture description (3GPP TS 36.401 version 9.2.0). 9 (2010).

Tsai, C. W.; Lai, C. F.; Chao, H. C. and Vasilakos, A.V. Big data analytics: a survey. *Journal of Big Data* 2 no. 1 (2015): 21.

Wang, H. and Fapojuwo, A. O. A survey of enabling technologies of low power and long range machine-to-machine communications. *IEEE Communications Surveys & Tutorials* 19 no. 99 (2017): 2621–2639.

Yu, X. and Xue, Y. Smart grids: A cyber-physical systems perspective. *Proceedings of the IEEE* 104 no. 5 (2016): 1058–1070.

4

Big Data Optimization in Electric Power Systems: A Review

Iman Rahimi
Islamic Azad University

Abdollah Ahmadi
University of New South Wales

Ahmed F. Zobaa
Brunel University London

Ali Emrouznejad
Aston Business School

Shady H.E. Abdel Aleem
15th of May Higher Institute of Engineering

CONTENTS

4.1 Introduction 56
4.2 Background 56
4.3 Scientometric Analysis of Big Data 58
4.4 Big Data and Power Systems 64
    4.4.1 Big Data Optimization 64
    4.4.2 Application of Big Data in Power System Studies 65
4.5 Optimization Techniques Used in the Big Data Analysis 65
    4.5.1 Computational Method for Large-scale Unconstrained
         Optimization 66
    4.5.2 Numerical Approach for Nonsmooth Large-scale
         Optimization 67
    4.5.3 Big Data in Logistics Optimization 67
    4.5.4 Big Data Analytics Based on Convex and Nonconvex
         Optimization 68
    4.5.5 Metaheuristic Algorithms for Big Data Optimization 69
4.6 Conclusion 71
References 74

4.1 Introduction

There are different definitions of big data, and among them, the most common definition refers to three or five characteristics, called volume, velocity, variety, value, and veracity from (Laney, 2001). Volume could include terabyte, petabyte, exabyte, and zettabyte. Velocity describes how fast the data are retrieved and processed "Batch or streaming." Variety describes structured, semi-structured, and unstructured data (Laney, 2001; Zikopoulos and Eaton, 2011). Veracity explains the integrity and disorderliness of data, while value refers to how good is the "value" we derive from analyzing data? (Zicari et al., 2016).

Electrical power systems are networks of components arrayed to supply, transfer, and use electric power. In power system, models are used to predict and characterize operations. However, there is a necessity for powerful optimization algorithms for information processing to learn models as the size increase of data is becoming a global problem to solve large-scale optimization problems. Any optimization problem includes a real function to be maximized or minimized by systematically determination of input values from an allowed set of values. Richness and quantity of large data sets provide the potential to enhance statistical learning performance but require smart models that use the latent low-dimensional structure for effective data separation.

This chapter reviews the most recent scientific articles related to large and big data optimization in power systems. Optimization issues such as logistics in power systems and techniques including nonsmooth, nonconvex, and unconstrained large-scale optimization are presented. After a brief review of big data, scientometric analysis has been applied using keywords of "big data" and "power system." Besides, keywords analysis, network visualization, journal map, and bibliographic coupling analysis have been done to draw a path on big data works in power system problems. Also, the most common useful techniques in large-scale optimization in power system have been reviewed. At the end of this chapter, metaheuristic techniques in big data optimization are reviewed to show that many efforts have been involved in big data optimization in power system and systematically highlight some perspectives on big data optimization.

4.2 Background

Before starting the discussion about big data optimization, this section reviews the importance of big data projects. Analyzing the big data could release valuable information. Setting up a big data task is a challenge that requires many tasks and processes to be done alongside with data store.

To support a big data-based project, one first needs to analyze the data. There are specific data management tools for storing and analyzing large-scale data. Even in a simple project, there are several steps that must be performed. Figure 4.1 shows these steps that include data preparation, analysis, validation, collaboration, reporting, and access. They are briefed as follows:

Data preparation is the process of collecting, cleaning, and consolidating data into one file or data table to be used in the analysis.

Data analysis is the process of inspecting, cleansing, transforming, and modeling data to discover the useful information, draw conclusions, and support decision-making.

Data validation is the process of ensuring that data have undergone a kind of cleansing to ensure they have acceptable quality and are correct and useful.

Data collaboration means data visualization from all available different data sources while getting the data from the right people, in the right format, to be used in making effective decisions.

Data reporting is the process of collecting and submitting data to authorities augmented with statistics.

Data access typically refers to software and activities related to store, retrieve, or act on data housed in a database or other repository.

Big data analysis provides valuable opportunities to support decision-making in several areas, including education, manufacturing, and healthcare. For instance, big data analytics have helped yield healthcare improvements by providing personalized medicine and prescriptive analytics, while in manufacturing big data analysis provides an infrastructure for transparency in the

FIGURE 4.1
Process of data analysis.

TABLE 4.1

Application of Big Data

Area	Scholars
Healthcare	Huser and Cimino (2016), O'Donoghue and Herbert (2012), Mirkes et al. (2016), and Murdoch and Detsky (2014)
Manufacturing	Lee et al. (2014b), Li et al. (2015), and Lee et al. (2015)
Science	Guide (2014), Brumfiel (2011), Francis (2012), Swan (2014)
Technology	Tay (2010), Johnson (2010), Sullivan (2015), and Layton (2014)
Education	Manyika et al. (2011), Picciano (2012), and West (2012)
Media	Smith et al. (2012), Xu et al. (2016), Couldry and Turow (2014), and Burgess and Bruns (2012)

manufacturing industry, which is the ability to unravel uncertainties such as inconsistent component performance and availability. An example of big data in science is the NASA Center for Climate Simulation (NCCS) that stores 32 petabytes of climate observations and simulations on a discover super-computing cluster. Amazon, eBay, Facebook, and Google are some examples of the application of big data in today's technology. Also, McKinsey Global Institute is known as an entity that applies big data in educational aspects. Table 4.1 presents some areas of big data applications in different fields; additional examples can be found in Bihl et al. (2016).

4.3 Scientometric Analysis of Big Data

Every activity in the 21st century, such as financial transactions, research, sales and purchase, security, transport, automobile sectors, the Internet, and others, requires data. With the advances in technology and fast development of the Internet, people observe the extent of data and information that enable access to vast amounts of data in a simple manner. However, this also needs a large amount of data with suitable storage capacity to host them. Nowadays, data manipulation techniques and computational capacities are some of the issues arising from big data, in which the classic technologies are not able to deal with them. Many researchers are working to resolve these problems in various areas such as health, economic, business, physics, and social sectors.

To highlight and show the importance of big data in today's power systems, scientometric technique and social network analysis (SNA) are used in the literature review. Recently, these techniques have become widespread because they facilitate understanding of some dynamical features such as collaboration among scholars (De Stefano et al., 2011; Emrouznejad and Marra, 2016; Lee et al., 2014a). Simply, they are known as strategic intelligence tools for the control of an emerging technology (Rotolo et al., 2014).

Scientometric is a key enabler that observes scientific publications to explore the structure and growth of a specific science using some quantitative measures of scientific information, as the number of scientific articles published in a given period, their citation impact, etc. (Rajendran et al., 2011). The main idea is to visualize data on behalf of a principal subject area to signify the whole activities in scientific output. The scientometric mapping technique is used to find the most common keywords that were used in recent research articles. For this aim, the title "large-scale power system" is searched in SCOPUS database which recalled about 1,107 scientific articles. Figure 4.2 presents the distribution of these papers from the 1970s.

Figure 4.3 presents a cognitive map where the size of the node is the equivalent number of publications on the considered term. Links among disciplines are shown by a lie whose density is proportional to the level of which two topics were being used in one article. The color of an item is managed by the cluster to which it belongs.

The most commonly used keywords (ten keywords) and their number of occurrences are given in Table 4.2. The objective of keyword analysis is to analyze the terms in a good accuracy. The process mainly depends on brainstorming to find the keywords which still have a high number of searches.

Figure 4.4 presents a different visualization of a country map that indicates collaboration among authors from different countries by lines. Authors from around 101 countries have collaborated in developing articles in big data and power systems. Figure 4.4 shows that China is the most active country in the power system field, then, the USA and Japan are at the second and third stages, respectively. Table 4.3 presents rank of the top five organizations,

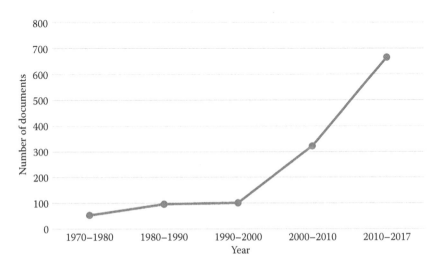

FIGURE 4.2
A number of publications on "large-scale" power system.

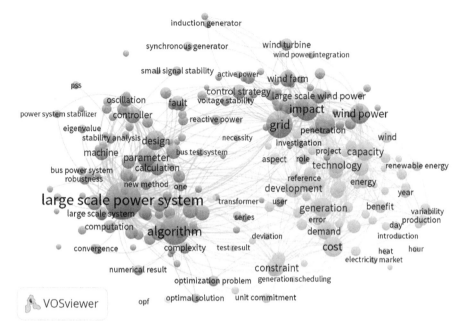

FIGURE 4.3
Cognitive map (keyword search based on co-occurrences).

TABLE 4.2

The Most Commonly Used Keywords in Big Data
Optimization Literature

No.	Keyword	Occurrences
1	Large-scale power system	399
2	Algorithm	248
3	Grid	211
4	Technique	166
5	Impact	152
6	Wind power	119
7	Cost	116
8	Integration	114
9	Capacity	110
10	Development	107

which have been addressed in affiliations of authors, with respect to the number of documents and citations.

Also, collaboration among authors has been analyzed. Figure 4.5 presents coauthor collaborations to display the robust and fruitful connections among collaborating researchers. The links across the networks in Figure 4.5

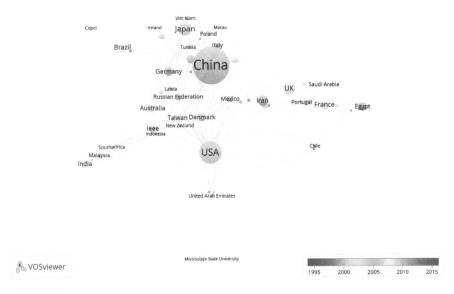

FIGURE 4.4
Network visualization (collaboration between countries).

TABLE 4.3

Rank of the Top Five Organizations by Number of Documents

No.	Organization	Number of Documents	Number of Citations
1	China Electric Power Research Institute	9	110
2	North China Electric Power University	7	247
3	Tsinghua University	4	36
4	University of Queensland	4	40
5	Brunel University	4	6

show the scientific communities involved in research on power systems and large-scale problems.

Figures 4.6 and 4.7 show network visualization and density map of the active journals in power system and large-scale problems based on citation analysis. Figure 4.6 presents the journals aggregated by density. The color shows the density, where the red color indicates a high density of a journal, while the blue color indicates the low-density journals. The right side of Figure 4.7 shows the densest area, occupied by journals dealing with the power system. The most frequent hosting sites are *IEEE Transaction on Power System, Applied Mechanics and Material, Power System Protection and Control, Automation of Electric Power System, IEEE Power and Energy Society General*

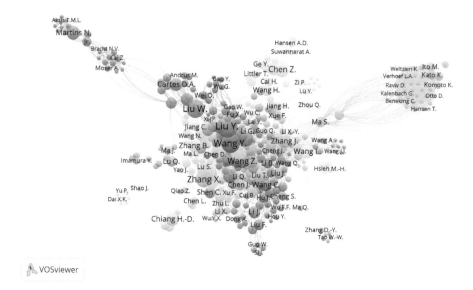

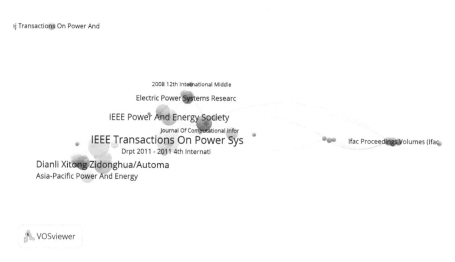

FIGURE 4.5
Scientific community (coauthor) working on the large-scale power system.

FIGURE 4.6
Journal map (title) based on citation analysis.

Meeting, International Journal of Electrical Power and Energy Systems, and *Proceedings of the Chinese Society of Electrical Engineering.*

Figure 4.8 shows different analysis (co-citation) of cited journals which possesses a minimum of ten citations for each source, and this leads to 152 sources with co-citation links.

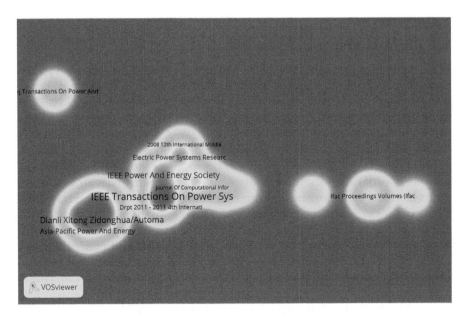

FIGURE 4.7
Density map (Journal title) based on citation analysis.

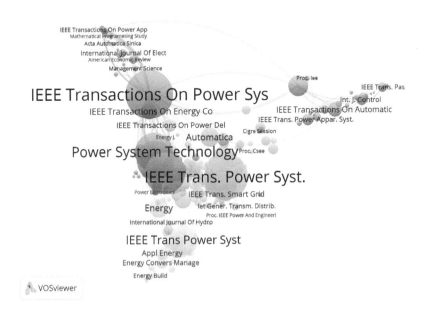

FIGURE 4.8
Network visualization (co-citation analysis).

4.4 Big Data and Power Systems

There are many large-scale optimization problems in power system, especially in the cases which consider the uncertainty of input parameters (Ahmadi et al., 2013, 2016; Charwand et al., 2015a,b; Esmaeel Nezhad et al., 2015; Mavalizadeh and Ahmadi, 2014; Sharafi Masouleh et al., 2016). Various researchers (cf. Ahmadi et al., 2014; Charwand et al., 2015a,b; Esmaeel Nezhad et al., 2015) consider the optimal operation of an electrical energy retailers. Ahmadi et al. (2016) propose a stochastic programing for the optimal operation for a distribution company. Mavalizadeh and Ahmadi (2014) consider emission and security for generation and transmission expansion planning. Ahmadi et al. (2011) and Sharafi Masouleh et al. (2016) use a mixed integer linear model for the optimal operation of hydro generation units. Moghimi et al., (2013) and Ghaikolaei et al. (2012) investigate the effects of distributed energy resources in the short-term optimal operation of power systems. Aghaei et al. (2015a,b), Esmaeily et al. (2017), and Karami et al. (2013) suggest using a Roulette wheel mechanism and lattice Monte Carlo simulation methods for modeling of uncertainties in hydrothermal scheduling problem. Ahmadi et al. (2014, 2016), Charwand et al. (a,b), Esmaeel Nezhad et al. (2015), Mavalizadeh and Ahmadi (2014), and Sharafi Masouleh et al. (2016) have many integer variables; for example, Aghaei et al. (2015a) report that the last case study has 3,841,392 variables, 1,610,808 discrete variables, and 4,712,112 equations. This example shows that the numbers of variables and equations are high. In the following sections, the background of big data in power systems is presented along with applications and the most common approaches in big data optimization in power systems.

4.4.1 Big Data Optimization

Big data optimization is one of the important issues in big data areas that have been widely arisen with many challenges such as privacy, size of data, and data management (Zicari et al., 2016). Social network science, machine learning, and biology are instances of many noticeable application fields where it is easy to formulate optimization problems with millions of variables. However, there is a necessity for powerful optimization algorithms for information processing to learn models as the size increase of data is becoming a global problem to solve large-scale optimization problems. Classical optimization algorithms are not planned to measure to cases of this size; new methods are required. Some examples of mathematical optimization in big data include logistics and supply chain issues (Brouer et al., 2016; Gunasekaran et al., 2017; Kache et al., 2017; Papadopoulos et al., 2017; Wu et al., 2017; Zhao et al., 2017), nonconvex optimization (Gong et al., 2016), unconstrained optimization (Babaie-Kafaki, 2016), and nonsmooth optimization

(Karmitsa, 2016). Big data optimization is usually taken into account in power systems research like management and scheduling, power dispatch, and energy demand.

4.4.2 Application of Big Data in Power System Studies

The use of big data has increased in several ways so that private companies and governments are investing billions of dollars in data management and analysis (Cukier, 2010). In power systems, data could be gathered from different sources such as renewables like solar and wind energies or other portions of energy technologies such as gas and fuel. In this regard, there are several applications of big data in energy domain that could be surveyed as renewables data use in biomass energy (Paro and Fadigas, 2011), marine energy (MacGillivray et al., 2014; Wood et al., 2010), wind energy (Billinton and Gao, 2008; Kaldellis, 2002), and energy consumption (Kung and Wang, 2015), or may consider energy-demand response such as power demand (Liu et al., 2013), and storage capacity (Goyena et al., 2009), or could be analyzed as electric vehicles (EVs) (Jiang et al., 2016) such as driving pattern (Wu et al., 2010), energy management Su and Chow (2012), energy efficiency (Midlam-Mohler et al., 2009), driving range (Lee and Wu, 2015; Rahimi-Eichi et al., 2015), battery capacity (Shor, 1994), data quality (Zhang et al., 2015), and EV state (Soares et al., 2015).

Also, there are other challenges in storage and analysis of data, visualization, sharing, etc. (Boyd and Crawford, 2011). It is common to identify trends, spots of problems, and predictive analysis to gain useful information from data. However, it is a big challenge when the problem is faced with big data. So a feature that is necessary for a successful big data analytics system is the need to make the data "over-the-counter" for understanding and using the data satisfactorily. This is especially vital for "high-stakes data" used to make better decisions. Firms which are making plans for big data tend to propose methods that consume less expensive storage, and processing alternatives, as well as tools to enhance data management. However, some of the significant challenges respondents cited to big data implementation are finding a staff to work in this domain and then training them while adjusting new methodologies for analytics and optimization.

4.5 Optimization Techniques Used in the Big Data Analysis

Traditional optimization methods could not be used to scale the large data size correctly; thus, new methods are critically needed. Optimization techniques in big data include several issues such as optimization big images, intelligent reduction, optimization based on Hadoop, and mathematical and metaheuristic optimization (Emrouznejad and Marra, 2016). There are numerous optimization methods that have been applied to power system operations. They are introduced as follows:

4.5.1 Computational Method for Large-scale Unconstrained Optimization

In some big data optimization programming, there are many variables resulting in a need for high memory. One of these applications is called unconstrained optimization which has broad application in engineering, industry, economic, and other fields. Unconstrained optimization also emerges from rewriting of constrained optimization by replacing some penalty terms in objective functions with some constraints. In this way, there is some application of unconstrained optimization method in power system problems (Zhu, 2015). While there are several approaches to dealing with unconstrained optimization, a conjugate gradient method is a useful method to solve large-scale cases (Babaie-Kafaki, 2016). Conjugate gradient techniques that were used for solving the linear system were suggested by Hestenes and Stiefel (1952). Required parameters for Hestenes–Stiefel (HS) method are introduced as follows:

$$\beta_k^{HS} = \frac{g_{k+1}^T}{d_k^T y_k} \quad K = 0, 1, \dots \tag{4.1}$$

where d_k is the search direction which is computed by inner products. This direction should be descent direction which means $g_k^T d_k < 0$, and $g_k = \nabla OF(x_k)$ where OF is a smooth nonlinear function that needs to be minimized, where $y_k = g_{k+1} - g_k$.

Regarding the mean-value theorem $\exists \zeta \in (0,1)$, we have

$$d_{k+1}^T y_k = d_{k+1}^T (g_{k+1} - g_k) = \alpha_k d_{k+1}^T \nabla^2 F(x_k + \zeta \alpha_k d_k) d_k \tag{4.2}$$

where α_k is a step length that is determined by the line search, and the condition $d_{k+1}^T y_k = 0$ can be considered as a conjugacy condition. Conjugate gradient methods include algorithms that are between Newton and steepest descent methods. Steepest descent method (Cauchy, 1847), Newton method (Sun and Yuan, 2006; Watkins, 2004), conjugate direction method (Babaie-Kafaki, 2016), and quasi-newton method (Sun and Yuan, 2006) are also applied for unconstrained optimization problems. Using the Hessian information, the techniques affect the direction of steepest descent. One of the weaknesses of the steepest descent technique was the slow convergence of the algorithm. In this regard, the method only needs the first-order derivatives, while the Newton method needs second-order derivative. These methods are broadly used for solving large-scale optimization problems.

4.5.2 Numerical Approach for Nonsmooth Large-scale Optimization

Definition of smooth functions arises from the first derivative (slope or gradient) at every point. In a graphical view, there is no abrupt in a smooth function of a single variable and also can be plotted as a single continuous; for example, the logistic loss $f(x) = \log(1 + \exp(-x))$ is a smooth function. In contrast, non-differentiable and discontinuous functions are classified as nonsmooth functions. Moreover, some functions with first derivatives also called non-differentiable. Graphs of non-differentiable functions may have abrupt bends, e.g., $f(x) = |x|$. These types of optimization are introduced as minimizing or maximizing which are broad in many applications such as economic (Outrata et al., 2013), engineering (Mistakidis and Stavroulakis, 2013), data analysis (Astorino and Fuduli, 2007; Astorino et al., 2008; Äyrämö, 2006), and control problem (Clarke et al., 2008). These problems are mostly large-scale. However, small-scale problems are also difficult to be solved (Karmitsa, 2016). The Boudle method is one of the techniques which could tackle large-scale nonsmooth optimization problem. There are two kinds of the bundle methods: limit memory bundle method (LMBM) and diagonal bundle method (D-bundle). Bundle method has also applied in different power system applications such as uncertainty (Bacaud et al., 2001), scheduling (Mezger and de Almeida, 2007; Zhang et al., 1999), and decomposition algorithms (Belloni et al., 2003; Borghetti et al., 2003). Some scholars have presented some works for nonsmooth functions (Attaviriyanupap et al., 2002; Dotta et al., 2009; Liu and Cai, 2005; Roy et al., 2010).

4.5.3 Big Data in Logistics Optimization

Logistics refers to actions which occur within the boundaries of single firms and supply chain mentions to networks of organizations which work together and coordinate their activities to deliver a product to market. Levels of the decision in the supply chain include three levels as illustrated in Figure 4.9 (Schmidt and Wilhelm, 2000). Decisions which determine the fleet size in marine logistics, for example, facility location and layout, belong to the strategic level. The logistics network may be possible to serve vast size of customers up to thousands of customers for a particular network. Operational level involves vehicle routing through transportation network, loading products, the landing of vessels, while tactical level involves production schedule and individual services (Brouer et al., 2016). However, Seaborn constitutes around 80% of transportation in the logistics network. In this case, network design problem is a primary planning problem in the logistics network. Regarding the demands which should be transported and selecting ports for servicing to supply chain decision makers wish to draw routes for their career to satisfy requirements of customers.

Sheu (2008) proposed a novel multi-objective optimization programming model to optimize operations in nuclear power generation (Taiwan nuclear

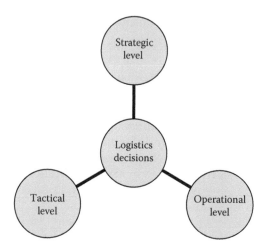

FIGURE 4.9
Different logistics decisions (Schmidt and Wilhelm, 2000).

power generation firm) and reduce waste logistics. The author has considered risk reduction in the formulation. The result depicts the improvement of performance from 7.41% to 18.37%, and risks were also reduced by 37.75%.

4.5.4 Big Data Analytics Based on Convex and Nonconvex Optimization

Mathematically, a single-objective minimizing (maximizing) optimization could be presented as follows:

$$\min(\max)OF(x)$$

$$s.t. g_i(x) \le 0, i = 1, \ldots, m \tag{4.3}$$

$$x \in D$$

where x is called a decision vector and D is the feasible region. OF is an objective function and g is constrained to function. Convexity condition for f, given D, holds the following condition:

$$OF(\lambda x_1 + (1 - \lambda)x_2) \le \lambda OF(x_1) + (1 - \lambda)OF(x_2) \quad \forall x1 \ne x2 \in D, \forall \lambda \in [0,1] \tag{4.4}$$

Equation (4.3) is called convex optimization problem if both functions OF and g are convex.

There is a possibility to find a global solution for Equation (4.3) if OF was convex. However, many real cases face the nonconvex optimization problem. In these cases, researchers try to find the local or global solution (Grossmann, 2014; Mistakidis and Stavroulakis, 2014). One of the relevant optimization problems in power system is known as the economic dispatch (ED). In the ED, the objective is defined allocating power demand among power plants in the most economic situation such that all operational constraints are satisfied. The cost function represents the quadratic fuel cost, and the valve-point effects cost which makes the objective function discontinuous, nonconvex. Selvakumar and Thanushkodi (2007) have applied a new particle swarm optimization (PSO) approach for nonconvex ED problem and suggested a new method in PSO based on the worst position of the particle and integrated it with local random search (LRS) and validated the proposed solution methodology with three ED tests. Their proposed algorithm shows significant improvement in convergence to the solution. Chaturvedi et al. (2009) used the PSO with time-varying acceleration coefficient in such a way that controls global and local search to achieve the global solution.

In many real applications, there are several objectives to be optimized. Multi-objective optimization usually includes conflict functions, in which improving one function leads to deterioration of the other one, so there is no single solution that can optimize all the functions together. In this case, researchers are looking for Pareto optimal solutions which are good compromising solutions. Equation (4.5) shows the following multi-objective problem:

$$\min(\max) \quad OF = \{OF_1(x), OF_2(x), \dots, OF_n(x)\},$$

$$s.t. \tag{4.5}$$

$$x \in D$$

Here, vector $x \in D$ is called Pareto solution to the problem (4.5) if there is no x^* such that $OF_i(x^*) \le OF_i(x)$ for any $i = 1, \dots, n$ and $\exists j (1 \le j \le n) : OF_j(x^*) < OF_j(x)$. If $OF(x^*) \le OF(x)$, it is said that x^* is a non-dominated solution. Guo et al. (2016) applied distributed optimization for a large-scale nonconvex transmission network. The authors applied spectral partitioning approach alongside the distributed optimization method, known as alternating direction method of multipliers (ADMM) to solve a nonconvex problem. In their work, they have shown that the solution found by ADMM is almost close to a local optimum.

4.5.5 Metaheuristic Algorithms for Big Data Optimization

Several new challenges have brought with the age of big data. Regarding optimization, researchers may face large-scale size problems, including hundreds, thousands, and even millions of variables. Several techniques have introduced and developed for tackling high-dimensional optimization problems. Among them, metaheuristic algorithms are known as efficient algorithms with high computing performance. Several scholars have used metaheuristic algorithms in power system (Camillo et al., 2016; Chen and Chang, 1995; Chiang, 2016; Lee and Yang, 1998; Rajesh et al., 2016). There are significant open research fields and issues for improvement. Among metaheuristic algorithms, evolutionary algorithms are known as a great powerful technique for continuous global optimization. However, increasing the number of variables resulting in deteriorating performance of the algorithm. There is a need for suitable approaches for dealing large-scale size problem to find global solutions to the optimization problems. Many scholars have attempted to face this difficulty (Beigvand et al., 2017; Chiou, 2007; Lin et al., 2017; Wang et al., 2010; Yan et al., 2004). An ED is a significant tool in power system operations, which schedules committed generating to meet demand in a point at a minimum cost (Beigvand et al., 2017). Beigvand et al. (2017) proposed hybridization of PSO and the Gravitational Search Algorithm (GSA) for a large-scale, nonconvex, nonsmooth, nonlinear, and noncontinuous combined heat and power dispatch. Summary of Beigvand et al. (2017) proposed algorithm is presented in Figure 4.10.

The authors have compared results with several optimization algorithms such as culture PSO (CPSO), modified PSO (MPSO), orthogonal teaching learning-based optimization (OTLBO), and teaching learning-based optimization (TLBO), GSA. Regarding robustness, the suggested method has better performance than other solution optimization methods. Moreover, the results show that hybrid algorithm has saved computational time significantly. Quality solution and the convergence speed of the hybrid algorithm possess superior performance than other optimization algorithms. Using of renewable energy has attracted the attention of power system planners across the world. Rajesh et al. (2016) applied differential evolution algorithm in a model of a solar plant to minimize both emission and cost. In the model, the data were gathered from demand and plants, and then the model is generated based on assumption. After several studies, the model is developed, and a solution methodology has been selected for the proposed

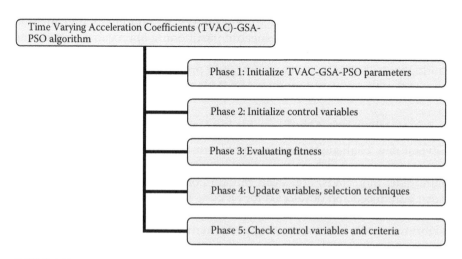

FIGURE 4.10
Phase classification for hybrid algorithm.

model. A sensitivity analysis was applied to the proposed model, and finally, the future power system model is generated with characteristics such as total cost, capacity additions, emission level. Naderi et al. (2017) proposed a fuzzy adaptive, comprehensive-learning PSO known as FAHCLPSO for the large-scale power dispatch optimization problem. Objective functions for the proposed algorithm include minimizing the active power transmission losses and improving the voltage profile of the system. The authors have validated the performance of their suggested algorithm with three different tests, including IEEE 30-bus, IEEE 118-bus, and IEEE 354-bus test systems. The authors have claimed that the proposed algorithm (FAHCLPSO) was the first applied for optimal reactive power dispatch. They have used fuzzy logic to enhance the searchability of the algorithm.

Tables 4.4 and 4.5 review classification of metaheuristic methods which have been carried out by scholars. Population-based approaches introduce most techniques and classified by evolutionary computations such as PSO, genetic algorithm (GA), Tabu search (TS), and ant colony optimization (ACO).

4.6 Conclusion

The chapter overviewed big data optimization issues in electric power systems. The scientific communities, distribution of publications, and collaboration among researchers around the world have been analyzed. The different types of big data optimization in power system have been discussed. Different types of complicated optimization problems in power systems were discussed. For this aim, factors such as nonlinearity of objective functions, number of variables, and nonsmooth functions were reviewed. One of the most difficulties dealing with these kinds of big data problems relates to the solution approach as addressed.

Because of the ongoing efforts in organizing smart grid infrastructure, the utility business is facing new challenges in dealing with big data and using them to improve decision-making. Big data in the electric power industry can be described in terms of volume, velocity, variety, veracity, value, or all the five terms. Usually, utilities do not handle data using an individual, consistent data management structure which makes ad-hoc use of the new decision-making packages needlessly complex. Although analysis of data is accessed through different data, if the data are not timed and spatial, unless they have a common data syntax and semantics for ease of use and if it is not fit for the uniform and common combination of the power system model, such analysis is perhaps not easy to implement. Moreover, one of the most challenging issues in power systems for decision makers arise from optimization problems.

TABLE 4.4

Literature of Metaheuristic Classification for Power System Problems

Metaheuristic									
Population							Local Search		
Naturally Inspired							Trajectory	Implicit	No Memory
		Implicit	Explicit	Direct					
Genetic Algorithm	Ant Colony	Evolutionary Programming	Differential Evolution	Simulated Annealing	PSO		Tabu Search	Scatter Search	Stochastic Local Search
Chiang (2005), Gerbex et al. (2001), and Walters and Sheble (1994)	Hou et al. (2002), Hou et al. (2004), and Niu et al. (2010)	Khatod et al. (2014), Tsai and Hsu (2010), Chung et al. (2010), and Yang et al. (1996)	Lakshminarasimman and Subramanian (2006), Liang et al. (2007), Su and Lee (2004), and Sayah and Zehar (2008)	Abido (2000), Zhuang and Galiana (1990), and Basu (2005)	Surendra and Parthasarathy (2014), Syahputra and Soesanti (2015), and Pan and Das (2016)		Lin et al. (2002), Abido (1999), and Mori and Goto (2000)	E Silva et al. (2014), Mori and Shimomugi (2007), and Mizutani et al. (2005)	Das and Patvardhan (1998), Das and Patvardhan (2002), and Hoos and Stützle (2004)

TABLE 4.5

Literature of Metaheuristic Classification for Power System Problems

Metaheuristic								
Population						Trajectory	Local Search	
Naturally Inspired		Implicit	Explicit	Direct			Implicit	No Memory
Genetic Algorithm	Ant Colony	Evolutionary Programming	Differential Evolution	Simulated Annealing	PSO	Tabu Search	Scatter Search	Stochastic Local Search
Panda and Yegireddy (2014), Apribowo and Hadi (2016), and Kaur et al. (2017)	Pothiya et al. (2010), Fetanat and Shafipour (2011), and Besheer and Adly (2012)	Wu and Ma (1995), Yuryevich and Wong (1999), and Lai (1998)	Cai et al. (2008), Shaheen et al. (2011), and Wang et al. (2009)	Abido (2000), Romero et al. (1995), and Lyden and Haque (2016)	Ahila et al. (2015), Abderrezek et al. (2016), Rouhi and Effatnejad (2015), Park et al. (2005), Niknam (2010), and Park et al. (2004)	Ramírez-Rosado and Domínguez-Navarro (2006), Katsigiannis et al. (2016), and Asadpour et al. (2015)	Habibi et al. (2014), Castillo et al. (2007), and de Padua et al. (2015)	Hoos (1998), Newton et al. (2014), and Fukuta and Ito (2011)

In addition, the chapter shows a significant effort involved in large-scale handling optimization which led to several algorithms, including mathematical optimization and metaheuristic optimizations, which metaheuristic optimizations that have proven to be more accurate, more efficient, and faster than earlier algorithms. Issues such as logistics optimization as well as Nonsmooth, nonconvex, and unconstrained large-scale optimization are presented. Finally, some metaheuristic methods in large-scale power system optimization are reviewed.

References

Abderrezek, H., Aissa, A. & Harmas, M. N. 2016. Adaptive non-singular terminal synergetic power system control using PSO. In *2016 8th International Conference on Modelling, Identification and Control (ICMIC)*. IEEE, 449–454.

Abido, M. 1999. A novel approach to conventional power system stabilizer design using tabu search. *International Journal of Electrical Power & Energy Systems*, 21, 443–454.

Abido, M. 2000. Robust design of multimachine power system stabilizers using simulated annealing. *IEEE Transactions on Energy Conversion*, 15, 297–304.

Aghaei, J., Ahmadi, A., Rabiee, A., Agelidis, V. G., Muttaqi, K. M. & Shayanfar, H. A. 2015a. Uncertainty management in multiobjective hydro-thermal self-scheduling under emission considerations. *Applied Soft Computing*, 37, 737–750.

Aghaei, J., Karami, M., Muttaqi, K. M., Shayanfar, H. A. & Ahmadi, A. 2015b. MIP-based stochastic security-constrained daily hydrothermal generation scheduling. *IEEE Systems Journal*, 9, 615–628.

Ahila, R., Sadasivam, V. & Manimala, K. 2015. An integrated PSO for parameter determination and feature selection of ELM and its application in classification of power system disturbances. *Applied Soft Computing*, 32, 23–37.

Ahmadi, A., Aghaei, J. & Shayanfar, H. 2011. Stochastic self-scheduling of hydro units in joint energy and reserves markets. *2011 19th Iranian Conference on Electrical Engineering (ICEE)*. IEEE, 1–5.

Ahmadi, A., Charwand, M. & Aghaei, J. 2013. Risk-constrained optimal strategy for retailer forward contract portfolio. *International Journal of Electrical Power & Energy Systems*, 53, 704–714.

Ahmadi, A., Charwand, M., Siano, P., Nezhad, A. E., Sarno, D., Gitizadeh, M. & Raeisi, F. 2016. A novel two-stage stochastic programming model for uncertainty characterization in short-term optimal strategy for a distribution company. *Energy*, 117, 1–9.

Apribowo, C. H. B. & Hadi, S. P. 2016. Design of experiments to parameter setting in a genetic algorithm for optimal power flow with TCSC device. *International Conference of Industrial, Mechanical, Electrical, and Chemical Engineering (ICIMECE)*. IEEE, 73–78.

Asadpour, M., Ajhiri, F. A., Tazehkand, B. M. & Seyedarabi, M. H. 2015. Jointly RVM based channel estimation and PAPR reduction using modified Tabu search algorithm in power line communication systems. *Wireless Personal Communications*, 84, 2757–2775.

Astorino, A. & Fuduli, A. 2007. Nonsmooth optimization techniques for semisupervised classification. *IEEE Transactions on Pattern Analysis and Machine Intelligence*, 29, 2135–2142.

Astorino, A., Fuduli, A. & Gorgone, E. 2008. Non-smoothness in classification problems. *Optimisation Methods & Software*, 23, 675–688.

Attaviriyanupap, P., Kita, H., Tanaka, E. & Hasegawa, J. 2002. A hybrid EP and SQP for dynamic economic dispatch with nonsmooth fuel cost function. *IEEE Transactions on Power Systems*, 17, 411–416.

Äyrämö, S. 2006. Knowledge mining using robust clustering. PhD thesis, University of Jyväskylä, Jyväskylä.

Babaie-Kafaki, S. 2016. Computational approaches in large-scale unconstrained optimization. In A. Emrouznejad (Ed.), *Big Data Optimization: Recent Developments and Challenges*. Switzerland: Springer.

Bacaud, L., Lemaréchal, C., Renaud, A. & Sagastizábal, C. 2001. Bundle methods in stochastic optimal power management: A disaggregated approach using preconditioners. *Computational Optimization and Applications*, 20, 227–244.

Basu, M. 2005. A simulated annealing-based goal-attainment method for economic emission load dispatch of fixed head hydrothermal power systems. *International Journal of Electrical Power & Energy Systems*, 27, 147–154.

Beigvand, S. D., Abdi, H. & La Scala, M. 2017. Hybrid gravitational search algorithm-particle swarm optimization with time varying acceleration coefficients for large scale CHPED problem. *Energy*, 126, 841–854.

Belloni, A., Lima, A. D. S., Maceira, M. P. & Sagastizábal, C. A. 2003. Bundle relaxation and primal recovery in unit commitment problems. The Brazilian case. *Annals of Operations Research*, 120, 21–44.

Besheer, A. & Adly, M. 2012. Ant colony system based PI maximum power point tracking for stand alone photovoltaic system. *2012 IEEE International Conference on Industrial Technology (ICIT)*. IEEE, 693–698.

Bihl, T. J., Young II, W. A. & Weckman, G. R. 2016. Defining, understanding, and addressing big data. *International Journal of Business Analytics (IJBAN)*, 3(2), 1–32.

Billinton, R. & Gao, Y. 2008. Multistate wind energy conversion system models for adequacy assessment of generating systems incorporating wind energy. *IEEE Transactions on Energy Conversion*, 23, 163–170.

Borghetti, A., Frangioni, A., Lacalandra, F. & Nucci, C. A. 2003. Lagrangian heuristics based on disaggregated bundle methods for hydrothermal unit commitment. *IEEE Transactions on Power Systems*, 18, 313–324.

Boyd, D. & Crawford, K. Six provocations for big data. *A Decade in Internet Time. Symposium on the Dynamics of the Internet and Society*. Oxford Internet Institute, Oxford, 2011.

Brouer, B. D., Karsten, C. V. & Pisinger, D. 2016. Big data optimization in maritime logistics. In A. Emrouznejad (Ed.), *Big Data Optimization: Recent Developments and Challenges*. Switzerland: Springer.

Brumfiel, G. 2011. Down the petabyte highway. *Nature*, 469, 282–284.

Burgess, J. & Bruns, A. 2012. Twitter archives and the challenges of "Big Social Data" for media and communication research. *M/C Journal*, 15(5).

Cai, H., Chung, C. & Wong, K. 2008. Application of differential evolution algorithm for transient stability constrained optimal power flow. *IEEE Transactions on Power Systems*, 23, 719–728.

Camillo, M. H., Fanucchi, R. Z., Romero, M. E., De Lima, T. W., Da Silva Soares, A., Delbem, A. C. B., Marques, L. T., Maciel, C. D. & London, J. B. A. 2016. Combining exhaustive search and multi-objective evolutionary algorithm for service restoration in large-scale distribution systems. *Electric Power Systems Research*, 134, 1–8.

Castillo, A., Ortiz, J. J., Perusquía, R., Hernandez, J. & Montes, J. 2007. Control rod pattern design using scatter search. *International Congress On Advances In Nuclear Power Plants*, 13–18 May. 2007, p. 6.

Cauchy, A. 1847. Méthode générale pour la résolution des systemes d'équations simultanées. *Comptes Rendus Mathematique Academie des Sciences, Paris*, 25, 536–538.

Charwand, M., Ahmadi, A., Heidari, A. R. & Nezhad, A. E. 2015a. Benders decomposition and normal boundary intersection method for multiobjective decision making framework for an electricity retailer in energy markets. *IEEE Systems Journal*, 9, 1475–1484.

Charwand, M., Ahmadi, A., Siano, P., Dargahi, V. & Sarno, D. 2015b. Exploring the trade-off between competing objectives for electricity energy retailers through a novel multi-objective framework. *Energy Conversion and Management*, 91, 12–18.

Chaturvedi, K. T., Pandit, M. & Srivastava, L. 2009. Particle swarm optimization with time varying acceleration coefficients for non-convex economic power dispatch. *International Journal of Electrical Power & Energy Systems*, 31, 249–257.

Chen, P.-H. & Chang, H.-C. 1995. Large-scale economic dispatch by genetic algorithm. *IEEE Transactions on Power Systems*, 10, 1919–1926.

Chiang, C.-L. 2005. Improved genetic algorithm for power economic dispatch of units with valve-point effects and multiple fuels. *IEEE Transactions on Power Systems*, 20, 1690–1699.

Chiang, C. L. 2016. An optimal economic dispatch algorithm for large scale power systems with cogeneration units. *European Journal of Engineering Research and Science*, 1, 10–16.

Chiou, J.-P. 2007. Variable scaling hybrid differential evolution for large-scale economic dispatch problems. *Electric Power Systems Research*, 77, 212–218.

Chung, C., Liang, C., Wong, K. & Duan, X. 2010. Hybrid algorithm of differential evolution and evolutionary programming for optimal reactive power flow. *IET Generation, Transmission & Distribution*, 4, 84–94.

Clarke, F. H., Ledyaev, Y. S., Stern, R. J. & Wolenski, P. R. 2008. *Nonsmooth Analysis and Control Theory*. Springer Science & Business Media, New York.

Couldry, N. & Turow, J. 2014. Advertising, big data and the clearance of the public realm: Marketers' new approaches to the content subsidy. *International Journal of Communication*, 8, 1710–1726.

Cukier, K. 2010. Data, data everywhere: A special report on managing information. *Economist Newspaper*.

Das, D. B. & Patvardhan, C. 1998. New multi-objective stochastic search technique for economic load dispatch. *IEE Proceedings-Generation, Transmission and Distribution*, 145, 747–752.

Das, D. B. & Patvardhan, C. 2002. Reactive power dispatch with a hybrid stochastic search technique. *International Journal of Electrical Power & Energy Systems*, 24, 731–736.

de Padua, S. G. B., Mantovani, J. R. S. & Cossi, A. M. 2015. Planning Medium-Voltage Electric Power Distribution Systems through a Scatter Search Algorithm. *IEEE Latin America Transactions*, 13(8), 2637–2645.

de Stefano, D., Giordano, G. & Vitale, M. P. 2011. Issues in the analysis of co-authorship networks. *Quality & Quantity*, 45, 1091–1107.

Dotta, D., E Silva, A. S. & Decker, I. C. 2009. Design of power system controllers by nonsmooth, nonconvex optimization. *Power & Energy Society General Meeting, 2009. PES'09. IEEE.* IEEE, 1–7.

E Silva, M. D. A. C., Klein, C. E., Mariani, V. C. & Dos Santos Coelho, L. 2014. Multiobjective scatter search approach with new combination scheme applied to solve environmental/economic dispatch problem. *Energy*, 53, 14–21.

Emrouznejad, A. & Marra, M. 2016. Big data: Who, what and where? Social, cognitive and journals map of big data publications with focus on optimization. In A. Emrouznejad (Ed.), *Big Data Optimization: Recent Developments and Challenges*. Switzerland: Springer.

Esmaeel Nezhad, A., Ahmadi, A., Javadi, M. S. & Janghorbani, M. 2015. Multi-objective decision-making framework for an electricity retailer in energy markets using lexicographic optimization and augmented epsilon-constraint. *International Transactions on Electrical Energy Systems*, 25, 3660–3680.

Esmaeily, A., Ahmadi, A., Raeisi, F., Ahmadi, M. R., Nezhad, A. E. & Janghorbani, M. 2017. Evaluating the effectiveness of mixed-integer linear programming for day-ahead hydro-thermal self-scheduling considering price uncertainty and forced outage rate. *Energy*, 122, 182–194.

Fetanat, A. & Shafipour, G. 2011. Generation maintenance scheduling in power systems using ant colony optimization for continuous domains based 0–1 integer programming. *Expert Systems with Applications*, 38, 9729–9735.

Francis, M. 2012. Future telescope array drives development of exabyte processing. *Ars Technica*.

Fukuta, N. & Ito, T. 2011. Toward combinatorial auction-based better electric power allocation on sustainable electric power systems. *2011 IEEE 13th Conference on Commerce and Enterprise Computing (CEC)*. IEEE, 392–399.

Gerbex, S., Cherkaoui, R. & Germond, A. J. 2001. Optimal location of multi-type FACTS devices in a power system by means of genetic algorithms. *IEEE Transactions on Power Systems*, 16, 537–544.

Ghadikolaei, H. M., Ahmadi, A., Aghaei, J. & Najafi, M. 2012. Risk constrained self-scheduling of hydro/wind units for short term electricity markets considering intermittency and uncertainty. *Renewable and Sustainable Energy Reviews*, 16, 4734–4744.

Gong, M., Cai, Q., Ma, L. & Jiao, L. 2016. Big network analytics based on nonconvex optimization. In A. Emrouznejad (Ed.), *Big Data Optimization: Recent Developments and Challenges*. Switzerland: Springer.

Goyena, S. G., Sádaba, Ó. A. & Acciona, S. 2009. Sizing and analysis of big scale and isolated electric systems based on renewable sources with energy storage. *2009 IEEE PES/IAS Conference on Sustainable Alternative Energy (SAE)*. IEEE, 1–7.

Grossmann, I. E. 2014. *Global Optimization in Engineering Design*. Springer Science & Business Media, New York.

Guide, L. 2014. English version: A collection of facts and figures about the Large Hadron Collider (LHC) in the form of questions and answers, CERN-Brochure-2009-003-Eng. *CERN*. English Version.

Gunasekaran, A., Papadopoulos, T., Dubey, R., Wamba, S. F., Childe, S. J., Hazen, B. & Akter, S. 2017. Big data and predictive analytics for supply chain and organizational performance. *Journal of Business Research*, 70, 308–317.

Guo, J., Hug, G. & Tonguz, O. K. 2016. A case for non-convex distributed optimization in large-scale power systems. *IEEE Transactions on Power Systems*, 32(5), 3842–3851.

Habibi, M., Rashidinejad, M., Zeinaddini-Meymand, M. & Fadainejad, R. 2014. An efficient scatter search algorithm to solve transmission expansion planning problem using a new load shedding index. *International Transactions on Electrical Energy Systems*, 24, 153–165.

Hestenes, M. R. & Stiefel, E. 1952. Methods of Conjugate Gradients for Solving Linear Systems. NBS, *Journal of Research of the National Bureau of Standards*, 49.

Hoos, H. H. 1998. *Stochastic Local Search-methods, Models, Applications*. Ios Press, Amsterdam, The Netherlands.

Hoos, H. H. & Stützle, T. 2004. *Stochastic Local Search: Foundations and Applications*. Morgan Kaufmann, San Francisco.

Hou, Y.-H., Wu, Y.-W., Lu, L.-J. & Xiong, X.-Y. 2002. Generalized ant colony optimization for economic dispatch of power systems. *International Conference on Power System Technology, 2002. Proceedings. PowerCon 2002*. IEEE, 225–229.

Hou, Y.-H., Xiong, X.-Y., Wu, Y.-W. & Lu, L.-J. 2004. Economic dispatch of power systems based on generalized ant colony optimization method. *Proceedings of the Csee*, 3, 014.

Huser, V. & Cimino, J. J. 2016. Impending challenges for the use of big data. *International Journal of Radiation Oncology Biology Physics*, 95, 890–894.

Jiang, H., Wang, K., Wang, Y., Gao, M. & Zhang, Y. 2016. Energy big data: A survey. *IEEE Access*, 4, 3844–3861.

Johnson, R. 2010. Scaling facebook to 500 million users and beyond. Retrieved May 4, 2014.

Kache, F., Kache, F., Seuring, S. & Seuring, S. 2017. Challenges and opportunities of digital information at the intersection of big data analytics and supply chain management. *International Journal of Operations & Production Management*, 37, 10–36.

Kaldellis, J. 2002. Optimum autonomous wind–power system sizing for remote consumers, using long-term wind speed data. *Applied Energy*, 71, 215–234.

Karami, M., Shayanfar, H., Aghaei, J. & Ahmadi, A. 2013. Scenario-based security-constrained hydrothermal coordination with volatile wind power generation. *Renewable and Sustainable Energy Reviews*, 28, 726–737.

Karmitsa, N. 2016. Numerical methods for large-scale nonsmooth optimization. In A. Emrouznejad (Ed.), *Big Data Optimization: Recent Developments and Challenges*. Switzerland: Springer.

Katsigiannis, Y., Kanellos, F. & Papaefthimiou, S. 2016. A software tool for capacity optimization of hybrid power systems including renewable energy technologies based on a hybrid genetic algorithm—Tabu search optimization methodology. *Energy Systems*, 7, 33–48.

Kaur, R., Krishnasamy, V., Muthusamy, K. & Chinnamuthan, P. 2017. A novel proton exchange membrane fuel cell based power conversion system for telecom supply with genetic algorithm assisted intelligent interfacing converter. *Energy Conversion and Management*, 136, 173–184.

Khatod, D. K., Pant, V. & Sharma, J. 2014. Evolutionary programming based optimal placement of renewable distributed generators. *IEEE Transactions on Power Systems*, 28, 683–695.

Kung, L. & Wang, H.-F. 2015. A recommender system for the optimal combination of energy resources with cost-benefit analysis. *2015 International Conference on Industrial Engineering and Operations Management (IEOM)*. IEEE, 1–10.

Lai, L. L. 1998. *Intelligent System Applications in Power Engineering: Evolutionary Programming and Neural Networks.* John Wiley & Sons, Inc, New York.

Lakshminarasimman, L. & Subramanian, S. 2006. Short-term scheduling of hydrothermal power system with cascaded reservoirs by using modified differential evolution. *IEE Proceedings-Generation, Transmission and Distribution*, 153, 693–700.

Laney, D. 2001. 3D data management: Controlling data volume, velocity and variety. *META Group Research Note*, 6, 70.

Layton, J. 2014. Amazon Technology. Retrieved from http://money.howstuffworks.com/amazon1.htm.

Lee, J.-D., Baek, C., Kim, H.-S. & Lee, J.-S. 2014a. Development pattern of the DEA research field: A social network analysis approach. *Journal of Productivity Analysis*, 41, 175–186.

Lee, J., Bagheri, B. & Kao, H.-A. 2015. A cyber-physical systems architecture for industry 4.0-based manufacturing systems. *Manufacturing Letters*, 3, 18–24.

Lee, C.-H. & Wu, C.-H. 2015. A novel big data modeling method for improving driving range estimation of EVs. *IEEE Access*, 3, 1980–1994.

Lee, J., Wu, F., Zhao, W., Ghaffari, M., Liao, L. & Siegel, D. 2014b. Prognostics and health management design for rotary machinery systems—Reviews, methodology and applications. *Mechanical Systems and Signal Processing*, 42, 314–334.

Lee, K. Y. & Yang, F. F. 1998. Optimal reactive power planning using evolutionary algorithms: A comparative study for evolutionary programming, evolutionary strategy, genetic algorithm, and linear programming. *IEEE Transactions on Power Systems*, 13, 101–108.

Li, J., Tao, F., Cheng, Y. & Zhao, L. 2015. Big data in product lifecycle management. *The International Journal of Advanced Manufacturing Technology*, 81, 667–684.

Liang, C., Chung, C., Wong, K. & Duan, X. 2007. Parallel optimal reactive power flow based on cooperative co-evolutionary differential evolution and power system decomposition. *IEEE Transactions on Power Systems*, 22, 249–257.

Lin, W.-M., Cheng, F.-S. & Tsay, M.-T. 2002. An improved tabu search for economic dispatch with multiple minima. *IEEE Transactions on Power Systems*, 17, 108–112.

Lin, S., Liu, M., Li, Q., Lu, W., Yan, Y. & Liu, C. 2017. Normalised normal constraint algorithm applied to multi-objective security-constrained optimal generation dispatch of large-scale power systems with wind farms and pumped-storage hydroelectric stations. *IET Generation, Transmission & Distribution*, 11(6), 1539–1548

Liu, D. & Cai, Y. 2005. Taguchi method for solving the economic dispatch problem with nonsmooth cost functions. *IEEE Transactions on Power Systems*, 20, 2006–2014.

Liu, Z., Wierman, A., Chen, Y., Razon, B. & Chen, N. 2013. Data center demand response: Avoiding the coincident peak via workload shifting and local generation. *Performance Evaluation*, 70, 770–791.

Lyden, S. & Haque, M. E. 2016. A simulated annealing global maximum power point tracking approach for PV modules under partial shading conditions. *IEEE Transactions on Power Electronics*, 31, 4171–4181.

Macgillivray, A., Jeffrey, H., Winskel, M. & Bryden, I. 2014. Innovation and cost reduction for marine renewable energy: A learning investment sensitivity analysis. *Technological Forecasting and Social Change*, 87, 108–124.

Manyika, J., Chui, M., Brown, B., Bughin, J., Dobbs, R., Roxburgh, C. & Byers, A. H. 2011. *Big Data: The Next Frontier for Innovation, Competition, and Productivity.* McKinsey Global Institute, New York.

Mavalizadeh, H. & Ahmadi, A. 2014. Hybrid expansion planning considering security and emission by augmented epsilon-constraint method. *International Journal of Electrical Power & Energy Systems,* 61, 90–100.

Mezger, A. J. & De Almeida, K. C. 2007. Short term hydrothermal scheduling with bilateral transactions via bundle method. *International Journal of Electrical Power & Energy Systems,* 29, 387–396.

Midlam-Mohler, S., Ewing, S., Marano, V., Guezennec, Y. & Rizzoni, G. 2009. PHEV fleet data collection and analysis. *Vehicle Power and Propulsion Conference, 2009. VPPC'09. IEEE.* IEEE, 1205–1210.

Mirkes, E., Coats, T. J., Levesley, J. & Gorban, A. 2016. Handling missing data in large healthcare dataset: A case study of unknown trauma outcomes. *Computers in Biology and Medicine,* 75, 203–216.

Mistakidis, E. S. & Stavroulakis, G. E. 2013. *Nonconvex Optimization in Mechanics: Algorithms, Heuristics and Engineering Applications by the FEM.* Springer Science + Business Media, Dordrecht.

Mizutani, A., Yukawa, T., Numa, K., Kuze, Y., Iizaka, T., Yamagishi, T., Matsui, T. & Fukuyama, Y. 2005. Improvement of input-output correlations of electric power load forecasting by scatter search. *Proceedings of the 13th International Conference on Intelligent Systems Application to Power Systems.* IEEE, p. 5

Moghimi, H., Ahmadi, A., Aghaei, J. & Rabiee, A. 2013. Stochastic techno-economic operation of power systems in the presence of distributed energy resources. *International Journal of Electrical Power & Energy Systems,* 45, 477–488.

Mori, H. & Goto, Y. 2000. A parallel tabu search based method for determining optimal allocation of FACTS in power systems. *Proceedings. PowerCon 2000. International Conference on Power System Technology, 2000.* IEEE, 1077–1082.

Mori, H. & Shimomugi, K. 2007. Transmission network expansion planning with scatter search. *IEEE International Conference on Systems, Man and Cybernetics, 2007. ISIC.* IEEE, 3749–3754.

Murdoch, T. B. & Detsky, A. S. 2014. The inevitable application of big data to health care. *JAMA,* 309, 1351–1352.

Naderi, E., Narimani, H., Fathi, M. & Narimani, M. R. 2017. A novel fuzzy adaptive configuration of particle swarm optimization to solve large-scale optimal reactive power dispatch. *Applied Soft Computing,* 53, pp. 441–456.

Newton, M. H., Pham, D. N., Tan, W. L., Portmann, M. & Sattar, A. 2014. Stochastic local search based channel assignment in wireless mesh networks. In Christian Schulte (ed.), *International Conference on Principles and Practice of Constraint Programming, Uppsala, Sweden.* Berlin: Springer, 832–847.

Niknam, T. 2010. A new fuzzy adaptive hybrid particle swarm optimization algorithm for non-linear, non-smooth and non-convex economic dispatch problem. *Applied Energy,* 87, 327–339.

Niu, D., Wang, Y. & Wu, D. D. 2010. Power load forecasting using support vector machine and ant colony optimization. *Expert Systems with Applications,* 37, 2531–2539.

O'Donoghue, J. & Herbert, J. 2012. Data management within mHealth environments: Patient sensors, mobile devices, and databases. *Journal of Data and Information Quality (JDIQ),* 4, 5.

Outrata, J., Kocvara, M. & Zowe, J. 2013. Nonsmooth *Approach* to *Optimization Problems* with *Equilibrium Constraints*: *Theory, Applications* and *Numerical Results*. Springer Science & Business Media, New York.

Pan, I. & Das, S. 2016. Fractional order fuzzy control of hybrid power system with renewable generation using chaotic PSO. *ISA Transactions*, 62, 19–29.

Panda, S. & Yegireddy, N. K. 2014. Automatic generation control of multi-area power system using multi-objective non-dominated sorting genetic algorithm-II. *International Journal of Electrical Power & Energy Systems*, 53, 54–64.

Papadopoulos, T., Gunasekaran, A., Dubey, R., Altay, N., Childe, S. J. & Fosso-Wamba, S. 2017. The role of big data in explaining disaster resilience in supply chains for sustainability. *Journal of Cleaner Production*, 142, 1108–1118.

Park, J.-B., Lee, K.-S., Shin, J.-R. & Lee, Y. 2004. Economic load dispatch for nonsmooth cost functions using particle swarm optimization. *Power Engineering Society General Meeting, 2003, IEEE*. IEEE, 938–944.

Park, J.-B., Lee, K.-S., Shin, J.-R. & Lee, K. Y. 2005. A particle swarm optimization for economic dispatch with nonsmooth cost functions. *IEEE Transactions on Power Systems*, 20, 34–42.

Paro, A. & Fadigas, E. 2011. A methodology for biomass cogeneration plants overall energy efficiency calculation and measurement—A basis for generators real time efficiency data disclosure. *Power Systems Conference and Exposition (PSCE), 2011 IEEE/PES*. IEEE, 1–7.

Picciano, A. G. 2012. The evolution of big data and learning analytics in american higher education. *Journal of Asynchronous Learning Networks*, 16, 9–20.

Pothiya, S., Ngamroo, I. & Kongprawechnon, W. 2010. Ant colony optimisation for economic dispatch problem with non-smooth cost functions. *International Journal of Electrical Power & Energy Systems*, 32, 478–487.

Rahimi-Eichi, H., Jeon, P. B., Chow, M.-Y. & Yeo, T.-J. 2015. Incorporating big data analysis in speed profile classification for range estimation. *2015 IEEE 13th International Conference on Industrial Informatics (INDIN)*. IEEE, 1290–1295.

Rajendran, P., Jeyshankar, R. & Elango, B. 2011. Scientometric analysis of contributions to Journal of Scientific and Industrial Research. *International Journal of Digital Library Services*, 1, 79–89.

Rajesh, K., Bhuvanesh, A., Kannan, S. & Thangaraj, C. 2016. Least cost generation expansion planning with solar power plant using differential evolution algorithm. *Renewable Energy*, 85, 677–686.

Ramírez-Rosado, I. J. & Domínguez-Navarro, J. A. 2006. New multiobjective tabu search algorithm for fuzzy optimal planning of power distribution systems. *IEEE Transactions on Power Systems*, 21, 224–234.

Romero, R., Gallego, R. & Monticelli, A. 1995. Transmission system expansion planning by simulated annealing. *Power Industry Computer Application Conference, 1995. Conference Proceedings. 1995 IEEE*. IEEE, 278–284.

Rotolo, D., Rafols, I., Hopkins, M. M. & Leydesdorff, L. 2014. Scientometric Mapping as a Strategic Intelligence Tool for the Governance of Emerging Technologies, SPRU (Science Policy Research Unit) Working Paper Series, p. 43.

Rouhi, F. & Effatnejad, R. 2015. Unit commitment in power system t by combination of Dynamic Programming (DP), Genetic Algorithm (GA) and Particle Swarm Optimization (PSO). *Indian Journal of Science and Technology*, 8, 134.

Roy, P., Ghoshal, S. & Thakur, S. 2010. Biogeography based optimization for multi-constraint optimal power flow with emission and non-smooth cost function. *Expert Systems with Applications*, 37, 8221–8228.

Sayah, S. & Zehar, K. 2008. Modified differential evolution algorithm for optimal power flow with non-smooth cost functions. *Energy Conversion and Management*, 49, 3036–3042.

Schmidt, G. & Wilhelm, W. E. 2000. Strategic, tactical and operational decisions in multi-national logistics networks: A review and discussion of modelling issues. *International Journal of Production Research*, 38, 1501–1524.

Selvakumar, A. I. & Thanushkodi, K. 2007. A new particle swarm optimization solution to nonconvex economic dispatch problems. *IEEE Transactions on Power Systems*, 22, 42–51.

Shaheen, H. I., Rashed, G. I. & Cheng, S. 2011. Optimal location and parameter setting of UPFC for enhancing power system security based on differential evolution algorithm. *International Journal of Electrical Power & Energy Systems*, 33, 94–105.

Sharafi Masouleh, M., Salehi, F., Raeisi, F., Saleh, M., Brahman, A. & Ahmadi, A. 2016. Mixed-integer programming of stochastic hydro self-scheduling problem in joint energy and reserves markets. *Electric Power Components and Systems*, 44, 752–762.

Sheu, J.-B. 2008. Green supply chain management, reverse logistics and nuclear power generation. *Transportation Research Part E: Logistics and Transportation Review*, 44, 19–46.

Shor, P. W. 1994. Algorithms for quantum computation: Discrete logarithms and factoring. *35th Annual Symposium on Foundations of Computer Science, 1994 Proceedings*. IEEE, 124–134.

Smith, M., Szongott, C., Henne, B. & Von Voigt, G. Big data privacy issues in public social media. *2012 6th IEEE International Conference on Digital Ecosystems Technologies (DEST)*. 2012. IEEE, 1–6.

Soares, J., Borges, N., Canizes, B. & Vale, Z. 2015. Probabilistic estimation of the state of electric vehicles for smart grid applications in big data context. *Power & Energy Society General Meeting, 2015 IEEE*. IEEE, 1–5.

Su, W. & Chow, M.-Y. 2012. Performance evaluation of an EDA-based large-scale plug-in hybrid electric vehicle charging algorithm. *IEEE Transactions on Smart Grid*, 3, 308–315.

Su, C.-T. & Lee, C.-S. 2004. Network reconfiguration of distribution systems using improved mixed-integer hybrid differential evolution. *IEEE Transactions on Power Delivery*, 18, 1022–1027.

Sullivan, D. 2015. Google still doing at least 1 trillion searches per year. March, 13.

Sun, W. & Yuan, Y.-X. 2006. *Optimization Theory and Methods: Nonlinear Programming*. Springer Science & Business Media, New York.

Surendra, U. & Parthasarathy, S. 2014. Optimal location of series FACTS device using PSO technique to reduce the losses and to enhance power transfer capability in a power system. In V. Sridhar, H. Seenappa Sheshadri, & M. C. Padma (Eds.), *Emerging Research in Electronics, Computer Science and Technology*. India: Springer.

Swan, M. 2014. The quantified self: Fundamental disruption in big data science and biological discovery. *Big Data*, 1, 85–99.

Syahputra, R. & Soesanti, I. 2015. Power system stabilizer model based on Fuzzy-PSO for improving power system stability. *Advanced Mechatronics. 2015 International Conference on Intelligent Manufacture, and Industrial Automation (ICAMIMIA)*. IEEE, 121–126.

Tay, L. 2013. Inside eBay's 90PB data warehouse, itNews, https://www.itnews.com. au/news/inside-ebays-90pb-data-warehouse-342615.

Tsai, M.-S. & Hsu, F.-Y. 2010. Application of grey correlation analysis in evolutionary programming for distribution system feeder reconfiguration. *IEEE Transactions on Power Systems*, 25, 1126–1134.

Walters, D. C. & Sheble, G. B. 1994. Genetic algorithm solution of economic dispatch with valve point loading. *IEEE Transactions on Power Systems*, 8, 1325–1332.

Wang, S.-K., Chiou, J.-P. & Liu, C.-W. 2009. Parameters tuning of power system stabilizers using improved ant direction hybrid differential evolution. *International Journal of Electrical Power & Energy Systems*, 31, 34–42.

Wang, Y., Li, B. & Weise, T. 2010. Estimation of distribution and differential evolution cooperation for large scale economic load dispatch optimization of power systems. *Information Sciences*, 180, 2405–2420.

Watkins, D. S. 2004. *Fundamentals of Matrix Computations*. John Wiley & Sons, New York.

West, D. M. 2012. Big data for education: Data mining, data analytics, and web dashboards. *Governance Studies at Brookings*, 4, 1–0.

Wood, R. J., Bahaj, A. S., Turnock, S. R., Wang, L. & Evans, M. 2010. Tribological design constraints of marine renewable energy systems. *Philosophical Transactions of the Royal Society of London A: Mathematical, Physical and Engineering Sciences*, 368, 4807–4827.

Wu, K.-J., Liao, C.-J., Tseng, M.-L., Lim, M. K., Hu, J. & Tan, K. 2017. Toward sustainability: Using big data to explore the decisive attributes of supply chain risks and uncertainties. *Journal of Cleaner Production*, 142, 663–676.

Wu, Q., Nielsen, A. H., Ostergaard, J., Cha, S. T., Marra, F., Chen, Y. & Træholt, C. 2010. Driving pattern analysis for electric vehicle (EV) grid integration study. *Innovative Smart Grid Technologies Conference Europe (ISGT Europe), 2010 IEEE PES*. IEEE, 1–6.

Wu, Q. H. & Ma, J. 1995. Power system optimal reactive power dispatch using evolutionary programming. *IEEE Transactions on Power Systems*, 10, 1243–1249.

Xu, Z., Liu, Y., Yen, N., Mei, L., Luo, X., Wei, X. & Hu, C. 2017. Crowdsourcing based description of urban emergency events using social media big data. *IEEE Transactions on Cloud Computing*.

Yan, W., Lu, S. & Yu, D. C. 2004. A novel optimal reactive power dispatch method based on an improved hybrid evolutionary programming technique. *IEEE Transactions on Power Systems*, 19, 913–918.

Yang, H.-T., Yang, P.-C. & Huang, C.-L. 1996. Evolutionary programming based economic dispatch for units with non-smooth fuel cost functions. *IEEE Transactions on Power Systems*, 11, 112–118.

Yuryevich, J. & Wong, K. P. 1999. Evolutionary programming based optimal power flow algorithm. *IEEE Transactions on Power Systems*, 14, 1245–1250.

Zhang, L., Chen, Y., Zhu, J., Pan, M., Sun, Z. & Wang, W. 2015. Data quality analysis and improved strategy research on operations management system for electric vehicles. *2015 5th International Conference on Electric Utility Deregulation and Restructuring and Power Technologies (DRPT)*. IEEE, 2715–2720.

Zhang, D., Luh, P. B. & Zhang, Y. 1999. A bundle method for hydrothermal scheduling. *IEEE Transactions on Power Systems*, 14, 1355–1361.

Zhao, R., Liu, Y., Zhang, N. & Huang, T. 2017. An optimization model for green supply chain management by using a big data analytic approach. *Journal of Cleaner Production*, 142, 1085–1097.

Zhu, J. 2015. *Optimization of Power System Operation*. John Wiley & Sons, New York.

Zhuang, F. & Galiana, F. 1990. Unit commitment by simulated annealing. *IEEE Transactions on Power Systems*, 5, 311–318.

Zicari, R. V., Rosselli, M., Ivanov, T., Korfiatis, N., Tolle, K., Niemann, R. & Reichenbach, C. 2016. Setting up a big data project: Challenges, opportunities, technologies and optimization. In A. Emrouznejad (Ed.), *Big Data Optimization: Recent Developments and Challenges*. Switzerland: Springer.

Zikopoulos, P. & Eaton, C. 2011. *Understanding Big Data: Analytics for Enterprise Class Hadoop and Streaming Data*. McGraw-Hill Osborne Media, New York.

5

Security Methods for Critical Infrastructure Communications

Ahmed F. Zobaa
Brunel University London

Trevor J. Bihl
Wright State University

CONTENTS

5.1   Introduction 86
5.2   Effects of Successful Communication System Threats 87
5.3   General Communication System Operations 87
5.4   Industrial Control Networks and Operations 89
    5.4.1   Industrial Control Network Operations and Components 89
    5.4.2   Commercial Technology Inroads into Industrial
    Control Networks 91
5.5   High-Level Communication System Threats 92
    5.5.1   Actor-Based Threats: Insider versus Outsider 92
    5.5.2   Device Property and Existential-Related Issues 93
    5.5.3   Host-Based Threats 94
    5.5.4   Physical versus Electronic Threats and Mitigation 94
    5.5.5   Supply-Chain-Related Threats and Mitigation 95
    5.5.6   Information Damage-Related Threats 95
    5.5.7   Stack-Based Exploitations 95
5.6   Cyber Threats and Security 96
    5.6.1   Component-Specific-Related Threats and Mitigation 97
    5.6.2   Software and Communication Threats and Mitigation 97
    5.6.3   Physical-Layer Threats and Security Measures 98
        5.6.3.1   Biometric-Like Security with Physical-Layer
        Security Measures 98
        5.6.3.2   Physically Traceable Objects 99
        5.6.3.3   Communication Signal Exploitation 100
5.7   Conclusions 101
References 101

5.1 Introduction

Critical infrastructure (CI) includes any systems and assets that are so vital that their destruction or disruption threatens lives, governments, economies, ecologies, or the social/political structure of nations (Luiijf & Klaver, 2004; Moteff & Parfomak, 2004). Thus, CI includes, but is not limited to, power grids, water and sewage, hospitals, and transportation systems (Luiijf & Klaver, 2004). To enable monitoring and control of CI systems, industrial control networks are often used (Galloway & Hancke, 2013). Industrial control networks, conceptualized in Figure 5.1, are systems that monitor and control physical devices. Conservatively, 80% of US electric power utilities employ industrial control networks for monitoring and control (Fernandez & Fernandez, 2005). Of interest in industrial control networks is preventing unauthorized access to CI systems and overall reliability of the networks.

Increasingly, commercial network technologies are being used in industrial control networks; this increases Internet pathways and cyber security risks. In many ways, extending the Internet of Things (IoT) to include CI components can be seen as logical since IoT-enabled devices can be used to monitor all components in a system, e.g., wireless-enabled structural health monitoring of bridges (Hu, Wang, & Ji, 2013). However, to be useful, communication networks used for CI need to balance performance, security, reliability, availability, and survivability (Ellison, Fisher, Linger, Lipson, & Longstaff, 1997; Snow, Varshney, & Malloy, 2000). Thus, beyond introducing vulnerabilities, security concerns can both limit user confidence in communication networks (Liao, Luo, Gurung, & Shi, 2015) and reduce their functionality (McMaster, 2003).

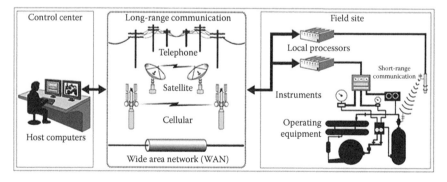

FIGURE 5.1
Conceptualization of an Industrial Control Network. (From Goverment Accountability Office (GAO), 2008.)

Communication security is only as strong as the weakest link, e.g., one insecure device in a large and otherwise security network can compromise the entire network (Yang, Luo, Ye, Lu, & Zhang, 2004). To secure networks, monitoring for anomalous behavior and vetting the identify of devices that aim to gain access to the network is critical. With the IoT expanding the volume and variety of devices connected to CI networks, the proliferation of communication devices and standards in CI applications thus presents security challenges. To understand how to vet the identity of communication devices, this chapter first reviews communications operations, then the types of devices used in industrial communication networks, the various threats to networks, and then the various security measures available.

5.2 Effects of Successful Communication System Threats

A variety of possible outcomes exist for successful CI communication system incidents. To this end, risk analysis of the various threats can be conducted concerning a communication system and the possible consequences if the threat occurs (Peltier, 2005). To evaluate what risks should be mitigated, security analysis can consider the likelihood of successful attack (L_{AS}) as a function of the threat (T), vulnerabilities (V), and target attractiveness (A_T) (Byres & Lowe, 2004). In conjunction with L_{AS}, of interest is also the consequence (C) of an attack (Byres & Lowe, 2004). While each communication system has various specific threats and possible consequences, the consequences can be generally binned as follows, where a malicious party could (Miller & Rowe, 2012): *distort* or modify files and information, *disrupt* access to the network, *disclose* information, *destroy* files or systems, or cause the *death* of humans. Additionally, some effects are *unknown* incidents, where the results and goals were not discovered by investigators (Miller & Rowe, 2012). With effective security analysis, the estimated financial, environmental, and health consequences of attacks can be estimated and used to allocate security resources (Byres & Lowe, 2004).

5.3 General Communication System Operations

In operation, industrial control networks are used for communication, monitoring, and controlling of devices and processes. For instance, instruments and operating equipment can record their states and transmit this as a message over the network. Similarly, an operator monitoring the equipment could send a message to change a state, e.g., opening a valve. However, to function, industrial control networks need a software and protocol framework to enable communications and routing of messages.

In general, to communicate over any network, first a software application (such as an operator clicking on a symbol for a valve he wishes to open) initiates the transmission of a data packet, which is the data or commands that are to be transmitted (Couch, 1993; Frenzel, 2013). This process is conceptualized in Figure 5.2 where the layers are conceptualized as the layers of the Open Systems Interconnection (OSI) model; consistent with Couch (1993; Frenzel, 2013). Table 5.1 provides general descriptions of each OSI layer. In general, all layers are software-related and indicate how data are handled; the exception is the physical layer which involves the physical components to transmit/receive data.

In Figure 5.2, as the packet proceeds through layers of software and hardware, more information is added to format the message, in the form of headers, addresses, etc. (Couch, 1993; Frenzel, 2013). These are added to describe the properties of devices, bit-level identification characteristics, communication properties, details for appropriate packet data handling, etc. (Couch, 1993; Frenzel, 2013). Once addresses, headers, and other details are added as data are conceptually passed through the OSI layers, the final message is transmitted over the communication medium (wired or wireless). Another device then receives the signal and the process is reversed to remove addresses and headers whereby it is determined how to handle and process the received data (Couch, 1993; Frenzel, 2013).

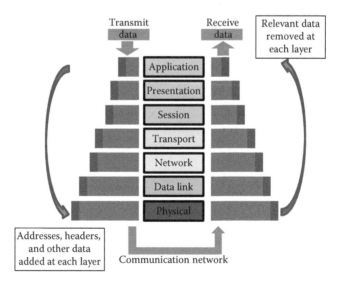

FIGURE 5.2
General digital communication operation. (From Bihl, 2015.)

TABLE 5.1

Communication Layers per the OSI Stack with Descriptions and Examples

	Data	Layer	Description	Example
Host layers	Data	Application	Software to access network	End user
		Presentation	Applies formatting to data, encrypts data, and facilitates application layer interaction.	Syntax and data manipulation
		Session	Interhost connections and session establishment	Synching
	Segments	Transport	Connection protocols	TCP and host-to-host
Media layers	Packets	Network	Determines physical path for data routing	Packets and routing
	Frames	Data link	Transfer of signal between nodes via physical devices	Frames and MAC addresses
	Bits	Physical	Signals, transmission, communication, and reception over a medium; physical components/devices	Cables, devices, physical mediums, and transmission methods

5.4 Industrial Control Networks and Operations

Industrial control networks operate by linking devices to operators via an infrastructure (wired, wireless, or a combination) (US Government Accountability Office, 2004; Slay & Miller, 2007). Human machine interfaces (HMI) feature prominently in industrial control networks and enable the presentation of interactive animations of devices and sensors, graphics of systems, and schematic diagrams to operators (Higgs, 2000; Gomez Gomez, 2005). Oversight and management for data acquisition in industrial control networks are provided by software layers including Distributed Control Systems (DCSs), Supervisory Control And Data Acquisition (SCADA) systems, Process Control Systems (PCS), and Cyber–Physical Systems (CPS) (Cárdenas, Amin, & Sastry, 2008; Galloway & Hancke, 2013). Appropriate and effective design of SCADA and DCS is key since industrial control networks are connected to physical equipment; they differ from commercial networks, e.g., wifi and Internet, by having high reliability requirements and the necessity of very short communication times but in small packets (Galloway & Hancke, 2013).

5.4.1 Industrial Control Network Operations and Components

Broadly, the structure in an industrial control network has four layers, as seen in Figure 5.1 and described in Table 5.2: processes and field equipment, devices, the station/substation of interest, and the enterprise (Dolezilek &

TABLE 5.2

Integration and Control (I&C) System Levels, per Dolezilek and Schweitzer (2000)

Level	Description	Example
Enterprise	Highest level, includes all end users who are inside or outside the substation.	Workstation at the corporate office.
Station/Substation	Third level, performs data acquisition and local input/output for the entire station.	Human machine interfaces, controller software, and decision support systems running on a local PC.
Device	Second level, contains PLCs and RTUs that collect and react to data.	Protective relays, meters, fault recorders, load tap changers, VAR controllers, RTUs, and PLCs
Process	Lowest level, connected to physical components for monitoring control.	Current transformers, voltage transformers, resistance thermal detectors, and transducers

Schweitzer, 2000; Slay & Miller, 2007; Schneider Electric, 2012). Processes include the industrial system itself and field equipment, such as sensors, instrumentation, and actuators (Schneider Electric, 2012). Devices include both Remote Telemetry Units (RTUs) and Programmable Logic Controllers (PLCs) (Kang & Robles, 2009; Galloway & Hancke, 2013). PLCs serve to control processes, perform digital and analog input/output, and provide control logic (Galloway & Hancke, 2013) and RTUs serve as sensor data collection devices (Slay & Miller, 2007; Galloway & Hancke, 2013). Due to ongoing developments in controllers, the function of RTUs and PLCs is frequently performed by an RTU/PLC device that can serve both functions as needed (Schneider Electric, 2012). The enterprise level includes the end user (Dolezilek & Schweitzer, 2000).

Bridging the gap between the station and enterprise level are the communication network and medium, host software, and the communication medium itself, as well as the protocols used to transmit data from RTUs and PLCs over the network (Galloway & Hancke, 2013). Finally, the host software layer includes software components, such as SCADA, whereby information is routed and presented effectively between clients, servers, and the field devices (Queiroz, Mahmood, Hu, Tari, & Yu, 2009; Galloway & Hancke, 2013). The client components refer to the end users, or operators, who monitor the system and the human–machine interface components (Daneels & Salter, 1999). Additional components can include firewalls and intrusion detection systems to protect the network from unauthorized access.

In practice, different RTU and PLC devices can be in use in a single installation and operate using different protocols; the languages used to exchange information (Daneels & Salter, 1999; Schneider Electric, 2012). Varying ages of devices in use exist in industrial control networks because outdated yet useful devices are rarely discarded while they continue to function correctly (Dolezilek & Schweitzer, 2000). Thus, one network could possibly contain many devices from many different manufacturers, and thus a variety of protocols can be found in use in any given industrial control network and stations. Additionally, proprietary versions of protocols can exist, making integration a further challenge (Schneider Electric, 2012); however, digital forensics investigations can aid in understanding proprietary protocol operations (Badenhop, Ramsey, Mullins, & Mailloux, 2016). Before being able to transmit over a network, one must understand and integrate effectively with the protocol. If proprietary protocols are used, this may require an operator to agree to a nondisclosure agreement (Badenhop, Ramsey, Mullins, & Mailloux, 2016), or to simply rely on the protocol to operate effectively. To facilitate communication, servers that aggregate information at the station level for communication over the network can generally handle multiple protocols (Daneels & Salter, 1999).

5.4.2 Commercial Technology Inroads into Industrial Control Networks

Commercial technology has made inroads into industrial control networks via two vectors: increased numbers of Internet pathways and increased use of Commercial Off the Shelf (COTS) communication devices in industrial control networks. Industrial control networks saw widespread use decades before the Internet (Robles & Choi, 2009). Since there were no initial pathways to commercial networks during this period, many industrial control networks regarded security as an afterthought (Cárdenas, Amin, & Sastry, 2008). However, widespread Internet connectivity has resulted in both indirect and direct pathways between it and industrial control networks (Patton et al., 2014); take, for instance, proposed industrial control interaction via direct Internet portals (Khatib, Dong, Qiu, & Liu, 2000) or cell phone applications (Ozdemir & Karacor, 2006). Additionally, recent advances in communication networks, such as Wireless Networks and the IoT, are increasingly finding use in CI systems (Jiang et al., 2014).

IoT advances and technologies, whereby communication abilities and links to everyday objects and devices (Wortmann & Flüchter, 2015), are increasingly finding use in CI systems to enable communication and monitoring of a wide number of devices. For example, smart grids might contain commercial wireless devices and protocols to enable meter or substation monitoring (Jiang et al., 2014). Traditionally, CI security focused on SCADA systems and protocols, while the IoT has expanded the number and types of devices and standards CI communication networks must consider (Mo et al., 2012). One type of IoT technology with increasing use in CI systems is the IEEE 802 standard subgroup (area networks) (IEEE, 2004). For instance, area networks have been, or been proposed for, used in CI, including the smart grid (Güngör et al., 2011), smart cities and e-government (Chang, Kannan, & Fellow, 2003;

Harmon, Castro-Leon, & Bhide, 2015), and CI applications such as hospitals (Cao, Leung, Chow, & Chan, 2009). However, notable security deficiencies exist in many commercial communication standards (cf. Melaragno, Bandara, Wijesekera, & Michael, 2012; Badenhop, Ramsey, Mullins, & Mailloux, 2016), thus including commercial communications devices introduces additional vectors for malicious agents to leverage.

Additionally, and naturally, connecting more and more CI devices through IoT advances results in big data concerns due to expanding volume, variety, and the velocity of signals transmitted. Due to the expanding variety and volume of devices in IoT CI implementations, future CI networks themselves have characteristics seen in the 3 V's (volume, variety, and velocity) of big data (Bihl, Young, & Weckman, 2016). Thus, monitoring logs and transmissions of communication devices to find threats can involve big data analytics due the massive amount of events logged (Samuelson, 2016; Gutierrez, Bauer, Boehmke, Saie, & Bihl, 2017).

5.5 High-Level Communication System Threats

Understanding cyber security involves understanding key characteristics of communication system threats. With an understanding of threats, one can develop and select appropriate security measures. Although a wide variety of threats exist, these can be grouped loosely by the approach taken, as conceptualized in the general taxonomy presented in Figure 5.3. In Figure 5.3, example threats include those related to the source (physical versus cyber), insider versus outsider (agent), etc.; this representation was adapted from Nawir, Amir, Yaakob, and Lynn (2016) by removing redundant groupings (information damage and access were synonymous) and introducing additional fields (e.g., supply chain related). A robust security approach mitigates these threats through a combination of both technological and nontechnological methods.

5.5.1 Actor-Based Threats: Insider versus Outsider

From a system perspective, threats emanate from inside or outside malicious actor(s), which dictate different courses of action to prevent and mitigate (Walton & Limited, 2006). Historically, most security breaches in corporations and industrial control networks threats were internal in nature; however, external (cyber) threats and breaches have become more common due to increasing Internet pathways within industrial control networks (Byres & Lowe, 2004).

Outsider threats are the work of hackers and malicious parties who wish to gain unauthorized access to a network and possibly disrupt its

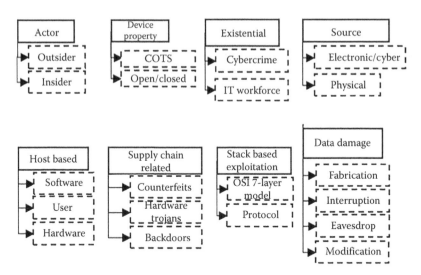

FIGURE 5.3
General taxonomy of communication system threats. (Adapted and Extended From Nawir, Amir, Yaakob, & Lynn, 2016.)

abilities (Walton & Limited, 2006). These threats inherently require technical approaches to mitigate and resolve (Walton & Limited, 2006). Conversely, insider threats are related to employees (past and present) and knowledgeable associates whose work is associated with the CI and communication system in question (Walton & Limited, 2006). Thus, insider threats are possibly immune to cyber security measures since malicious parties might know appropriate passwords, account details, etc., needed to achieve access.

Disaffected employees, and employees under the sway of blackmail, bribery, or ideology, way wish to disrupt or damage the network (Walton & Limited, 2006). Logically, one would desire to minimize insider threats completely and focus on outsider threats since insider threats are difficult to detect (Walton & Limited, 2006). However, three general approaches exist to detect and deter insider threats (Walton & Limited, 2006): (1) mitigating possible damages by compartmentalization; (2) early detection via authentication and auditing, and (3) proper management and ownership to reduce disaffection. Thus, a combination of proper security procedures, technology to find suspicious actions, and management all have roles in mitigating insider threats.

5.5.2 Device Property and Existential-Related Issues

Various properties of the communication devices related to proprietary and nonproprietary designs can be exploited. If a communication network uses COTS devices, then any system using these devices inherits their known and unknown vulnerabilities (Cárdenas, Amin, & Sastry, 2008). Prior to wide spread use of COTS devices in CI implementation, industrial control networks used mostly highly customized software and hardware components and thus had the advantage of "security by obscurity" (Stuttard, 2005). The open versus closed nature of protocol designs can also be related to vulnerabilities; closed/proprietary protocols have the advantage of "security by obscurity." Security by obscurity means that closed and proprietary protocols benefit from their obscurity, where malicious actors find difficulty learning the particulars to exploit. Open designs with public protocols do not benefit from security by obscurity; however, such networks are generally safer since security professionals can fix and augment security issues as they become known (Cárdenas, Amin, & Sastry, 2008).

Vulnerabilities also exist due to existential issues related to the expanding pool of skilled IT professionals throughout the world with the skills to attack communication systems (Cárdenas, Amin, & Sastry, 2008). Additionally, the amount of freely available cybercrime tools is expanding and available for use by even less-skilled malicious actors (Cárdenas, Amin, & Sastry, 2008). However, it should be noted that while a certain pool of skilled IT professionals can be malicious, it is also advantageous to security to find flaws and develop solutions (Rescorla, 2005).

5.5.3 Host-Based Threats

The host of the system can be compromised through various means as discussed by Nawir, Amir, Yaakob, and Lynn (2016). For instance, an authorized user might not effectively protect credentials and so a malicious actor could gain access to a network via those authentic credentials. Alternatively, an attacker could compromise software by overloading resource buffers or pushing devices to exhaustion. Finally, hardware can become compromised if malicious code is injected into it; for example, contact with infected flash drives was sufficient for the Stuxnet worm to infect computers which were not directly connected to the Internet (Chen & Abu-Nimeh, 2011).

5.5.4 Physical versus Electronic Threats and Mitigation

Outsider threats involve attacks on a communication network by parties who are remote and not directly connected to the organization that manages the network (Walton & Limited, 2006). Broadly, outsider threats can be physical, like the 2013 assault on PG&E's Metcalf transmission station (Smith, 2014), or electronic, like cyber-attacks on CI (Miller & Rowe, 2012). Electronic threats broadly include all other software and protocol exploitation methods. Here, the communication medium is used as a vector to infect, restrict access, or damage network operating conditions. Physical attacks on CI systems can be seen in the form of terrorists and criminals who gain in-person access to a site to physically attack it (Smith, 2014). While these attacks might not aim specifically at the communication system, damages and reduced capabilities could result. While a physical attack on infrastructure can be mitigated by site security, physical attacks via hardware trojans are stealthy in nature and could result from an insecure electronic supply chain.

5.5.5 Supply-Chain-Related Threats and Mitigation

While electronic/cyber is the primary security concern in communication systems, supply-chain concerns also exist since CI communication systems interact with physical objects and have many components, possibly at long distances from monitors. Outsourcing electronic production has introduced weaknesses in supply-chain security for electronics and introduced various issues (Jang-Jaccard & Nepal, 2014). Physical threats exist in the form of counterfeit electronics, which can fail quicker (Guin et al., 2014; Tehranipoor, 2015), integrated circuits (ICs) which have hardware trojans, integrity circuits, or malignant logic can compromise the security of a network (Di & Smith, 2007; Jang-Jaccard & Nepal, 2014), and compromised circuits can include backdoors to facilitate future attacks (Jang-Jaccard & Nepal, 2014). Collectively, robust physical security to limit unauthorized site access is necessary and includes secured IC supply chains (Karri, Rajendran, Rosenfeld, & Tehranipoor, 2010; Guin et al., 2014).

5.5.6 Information Damage-Related Threats

In addition to the specific effects discussed in Section 5.2, data damage actions are possible, as discussed by Nawir, Amir, Yaakob, and Lynn (2016). Once an actor or virus/worm has gained access to a network, various possible actions could happen to the data being collected. These data involve either the sensor readings from a substation and actions sent by an observer—so, data integrity is key to reliable operations. Threats exist to data in the interception of communications, whereby data might be monitored passively (eavesdropping) or even modified before it reaches its indented recipient. Data can also be fabricated to allow for situations where an attacker floods a system with data that show normal conditions while the actual system is in an out of control state. Additionally, an attacker could interrupt data, which might cause the communication network to shut down or merely replay the last observations received.

5.5.7 Stack-Based Exploitations

Additional threats exist due to the exploitation of protocol characteristics and different functionalities of communication operations. Knowledge of protocol specifics can lend itself to the exploitation of weaknesses in a given protocol. Additionally, the operations and characteristics associated with each of the OSI 7-layers are associated with particular weaknesses. From a security and device identification standpoint, different layers of the OSI stack also correspond to different information; for example, network-level encryption keys are "something you know," MAC-level MAC addresses are "something you have," and physical-layer characteristics are "something you are" (Ramsey, Temple, & Mullins, 2012). With this understanding, one can further understand attacks and issues per layer and determine appropriate cyber security measures.

5.6 Cyber Threats and Security

Focusing primarily on the electronic/cyber threats found in the Stack-based Exploitation and Data Damage threats in Figure 5.3 requires understanding the specific threats employed and protection methods. Table 5.3 presents various threats and protection measures in reference to the 7-layer OSI model of Table 6.1 with threats and protections per Nawir, Amir, Yaakob, and Lynn (2016). Broadly, we will characterize these threats and security measures as follows: component-specific, e.g., PLC security issues, physical-layer related, e.g., hardware threats, and then software and protocol-based, e.g., most of the issues found in Table 5.3.

TABLE 5.3

OSI 7 Layer Model with Example Threats and Protections Available per Layer

Layer	Threats	Protection
Application	Clock skewing, selective message forwarding, data aggregation distortion, and clone attacks	High-level firewalls
Presentation	SSL to tunnel HTTP attacks	Applications delivery platform (ADP)
Session	Hijacking	Packet analysis, encryption, and limiting packets
Transport	Renegotiation, port scans, DoS, misdirection, flooding, and de-synchronization	Handshake protocol
Network	False routing, packet replication, blackhole, wormhole, sinkhole, sybil, selective forwarding, HELLO flood, and acknowledgement spoofing	Firewalls and encryption keys
Data Link	MAC flooding, MAC spoofing, ARP cache poisoning, traffic manipulation, identity spook, collision, exhaustion, and unfairness	Intrusion detection/prevention systems (IDPS)
Physical	Device tampering, eavesdropping, jamming, and counterfeits	RF fingerprinting, PUFs, and COAs

5.6.1 Component-Specific-Related Threats and Mitigation

Security threats can exist due to weaknesses in specific components. Due to the large number of PLCs in an industrial control network, weaknesses found in these devices can be a critical vector for compromises to occur. Since PLCs monitor and control physical devices, realized threats related to PLCs can result in devices being driven out of safety margins and possibly to system damaging outcomes, e.g., Stuxnet (Chen & Abu-Nimeh, 2011). Threats to PLCs include worms that can infect and change memory values to arbitrary values resulting in a given PLC operating its control logic via incorrect values (Sandaruwan, Ranaweera, & Oleshchuk, 2013). Many PLCs also have forcing output functionalities which enable an operator to force an output to be a specific value; thus, any PLC with direct links to the Internet could be compromised if an attacker gains direct access (Sandaruwan, Ranaweera, & Oleshchuk, 2013). Finally, protocol exploitations, e.g., the malformed packets (Ultes-Nitsche & Yoo, 2004), can be used as a further software vector to PLC attacks (Sandaruwan, Ranaweera, & Oleshchuk, 2013).

5.6.2 Software and Communication Threats and Mitigation

A wide variety of communication devices and standards exist in CI implementations, including a variety of SCADA protocols, e.g., Modbus®, RP-570, Profibus, Conitel, IEC 61850, T101, IEC 60870-5-101 (104), DNP V3.0, ISO-TSAP, and UCA (Utility Communications Architecture) (Robles, Choi, & Kim, 2009). While not all of these protocols employ the OSI 7-layers as described in Table 6.1, the same broad operations are still performed per protocol operations and thus all are generally susceptible to the various attacks lists in Table 5.3. All of these standards are associated with various advantages and weaknesses. For example, the ISO-TSAP protocol used by many Siemens PLCs does not provide for data encryption (Sandaruwan, Ranaweera, & Oleshchuk, 2013). Limitations in specific protocols have also led to the development of secured versions of protocols, e.g., "Secure MODBUS" (Fovino, Carcano, Masera, & Trombetta, 2009).

Incorporating intrusion detection and prevention systems (IDPSs) into industrial control networks can mitigate MAC-related attacks and provide a log of events which violate access rules (Xing, Srinivasan, Jose, Li, & Cheng, 2010; Zhu & Sastry, 2010). However, IDPSs generally rely on coded rules, which are limited against new and novel attacks (Gutierrez, Bauer, Boehmke, Saie, & Bihl, 2017). A variety of network-based routing attacks exist and these can take the form of attackers flooding, or corrupting routing information or flooding the network with replicated packets to consume bandwidth and cause communication termination (Xing, Srinivasan, Jose, Li, & Cheng, 2010). Network attacks can be mitigated by routing access restrictions and detection methods that watch for false routing and other types of attacks (Xing, Srinivasan, Jose, Li, & Cheng, 2010). Higher level attacks can exist at the application level and influence the software used by the operator. For instance, clock skewing can desynchronize operations and cause communications to be unstable in protocols that require synchronization, e.g., wireless sensor networks operating under IEEE 802.11 (Xing, Srinivasan, Jose, Li, & Cheng, 2010). Authentication methods and data integrity approaches can be adopted to mitigate against these risks (Xing, Srinivasan, Jose, Li, & Cheng, 2010).

5.6.3 Physical-Layer Threats and Security Measures

At the physical layer, a variety of threats can exist. For instance, devices can be tampered with, and counterfeit ICs exist in the supply chain for many communication devices (Guajardo, Kumar, Schrijen, & Tuyls, 2008). Subsequently, various economic, security, and safety issues can exist; for example: counterfeit IC results in millions to billions of dollars in lost revenue to developers, security issues exist in that counterfeit ICs could be designed to learn operating keys, thereby allowing unauthorized access, and further issues exist for users since counterfeit ICs are more prone to failure (Guajardo, Kumar, Schrijen, & Tuyls, 2008). While software-based security often receives the majority of the emphasis, all software-based security is hackable as seen in Table 5.3. Thus, determining the authenticity of devices or individual ICs is also of interest for CI protection.

5.6.3.1 Biometric-Like Security with Physical-Layer Security Measures

Biometric security involves selecting using discriminating qualities that are *universal, distinct, permanent,* and *collectable* (Cobb, Garcia, Temple, Baldwin, & Kim, 2010). Biometric-like security for communication devices involves examining the intended and unintended communication and radiation are useful for device identification between disparate devices (Weng et al., 2005; Cobb, Laspe, Baldwin, Temple, & Kim, 2012). When devices from the same production run are considered, communication signal-fingerprinting approaches enable production-induced variations to be discriminable (Cobb, Laspe, Baldwin, Temple, & Kim, 2012). Physical-layer features and identification methods can be employed as an additional level of security whereby claimed identities are vetted for device identity authentication (Cobb, Laspe, Baldwin, Temple, & Kim, 2012). Physical-layer features aim to characterize communication devices due to production variations whereby minute signal differences can be used to discriminate between individual devices (Cobb, Laspe, Baldwin, Temple, & Kim, 2012).

Because physical-layer characteristics are associated with the intrinsic physics-based properties of devices, they provide inherent benefits in preventing spoofing attacks common with security at other OSI levels (Tomko, Rieser, & Buell, 2006). Desirable physical-layer characteristics are those that are identifiable and possess biometric-like qualities (see Jain, Ross, &

Prabhakar, 2004; Ryer, Bihl, Bauer, & Rogers, 2012) of *universality, distinctiveness, permanence,* and *collectability* (Cobb, Garcia, Temple, Baldwin, & Kim, 2010). Two general approaches of physical-layer security exist for this purpose: (1) adding physically traceable objects to devices (DeJean & Kirovski, 2007; Majzoobi, Koushanfar, & Potkonjak, 2009; Grau, Zeng, & Xiao, 2012) and (2) the exploiting inherent features present in device signals, e.g., through RF fingerprinting (Ellis & Serinken, 2001; Suski, Temple, Mendenhall, & Mills, 2008; Cobb, Garcia, Temple, Baldwin, & Kim, 2010; Scanlon, Kennedy, & Liu, 2010).

5.6.3.2 Physically Traceable Objects

Three identification methods have been proposed to verify the identity of communication devices using physically traceable objects: Radio Frequency Identification (RFID), Physical Unclonable Functions (PUFs), and RF Certificates of Authenticity (RF-COA). While there are various benefits to each approach, all are limited in their ability to be applied to equipment already in use.

RFID is a tracking technology which involves placing an identifier antenna "tag" on a device for tracking (Landt, 2005; Roberts, 2006). To identify devices, the RFID tag either actively emits (powered RFID tags) or emits only when scanned (unpowered RFID tags) (Grau, Zeng, & Xiao, 2012). Due to the ability to remotely track objects, RFID has seen extensive use in commercial and warehouse applications for products tracking (Landt, 2005). RFID does have known issues, including interference (Holland, Young, & Weckman, 2011), placement (RFID antennas must be located on each device), and type-level issues (multiple identical objects typically receive the same RFID tag).

Both PUFs and RF-COAs are an extension of the RFID process whereby uniquely identifiable components or antenna are added to an IC. While RFID tags operate at a type level, PUFs and RF-COAs operate at a serial-number level. PUFs include two techniques for authentication: (1) adding internal measurement circuitry to IC and (2) adding capacitive sensors on top of ICs in a grid form (Cobb, Laspe, Baldwin, Temple, & Kim, 2012). PUFs work by incorporating a randomized component to these augmentations, to ensure uniqueness (Cobb, Laspe, Baldwin, Temple, & Kim, 2012).

RF-COAs essentially take the RFID concept and make small, unique, and three-dimensional antennae using randomly shaped conductors and dielectric components which are placed onto ICs to create a uniquely identifiable RF signal (DeJean & Kirovski, 2007). In essence, RF-COAs combine PUFs and RFID into a single IC identification approach (Cobb, Laspe, Baldwin, Temple, & Kim, 2012).

Both PUFs and RF-COAs can be employed to ensure ICs are authentic in a similar way that product keys are used to ensure authorized installation of software (Guajardo, Kumar, Schrijen, & Tuyls, 2008). While PUFs can provide increased security, both PUF approaches require physical IC manipulations and thus are prohibitive for use with legacy devices. RF-COAs have similar, and obvious, impediments to their use on legacy devices, in addition to extra design considerations needed in the manufacturing and design process. Finally, RF-COAs are further limited in utility due to the existence of spoofing mechanisms (DeJean & Kirovski, 2007).

5.6.3.3 Communication Signal Exploitation

RF fingerprinting is the characterizing a communication device from minute differences in emanated signals to extract biometric-like features (Candore, Kocabas, & Koushanfar, 2009; Weber, Birkel, Collmann, & Engelbrecht, 2010). RF fingerprinting implies systematic signal collection, processing, sampling, statistical feature extraction methods, and classifier model development (Harmer, 2013). When considering intentional emissions, RF fingerprinting has been successful in discriminating inter-device variations, e.g., similar devices from different manufacturers (Klein, 2009), and intra-device variations, e.g., devices from the same manufacturer that differ only by serial number (Bihl, Bauer, & Temple, 2016).

After collecting signals, a region of interest, e.g., the preamble which should be consistent for a protocol, is isolated (Bihl, Bauer, & Temple, 2016). Instantaneous amplitude, phase, and frequency response are then computed for each region of interest (Bihl, Bauer, & Temple, 2016). These responses are then divided into bins, from which RF fingerprinting features are then extracted. The considered RF fingerprinting features are generally the second, third, and fourth mathematical moments (variance, skewness, and kurtosis), which are used to quantify distributional properties of the signal for identification (Thirukkonda, 2009; Cobb, 2011; Lohweg et al., 2013). Figure 5.4

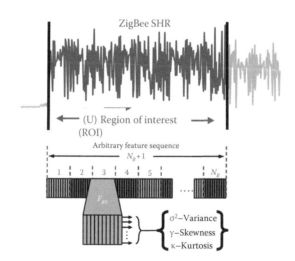

FIGURE 5.4
General RF fingerprinting process, example using with ZigBee signal. (From Bihl, 2015.)

presents a visualization of the RF-DNA fingerprints from sampled-time ZigBee preamble data.

Applicability of RF fingerprinting methods includes wireless and wired communications (cf. Carbino, Temple, & Bihl, 2015; Bihl, Bauer, & Temple, 2016). Recent advancements in adapting RF fingerprinting to wired communication include the works of Lopez, Temple, and Mullins (2014; Ross, Carbino, & Stone, 2017), both of which explored CI-related communication device discrimination. Outside of laboratory research, commercial devices have begun to provide physical-layer authentication ability, e.g., (PFP Cybersecurity, 2016).

5.7 Conclusions

To have a reliable industrial control network, one must consider effective security measures. Security primarily involves authenticating the identity of devices and operators, thus restricting unauthorized access to networks. Given the severity of intrusions in CI networks, preventing unauthorized access and limiting Internet pathways are necessary. However, the expansion of IoT into CI systems, e.g., the Smart Grid, precludes the ability to successfully rely on security through obscurity for industrial control networks, and thus effective cyber security strategy is necessary. Although much research and work exists in cyber security and authentication, these tend to be related to preventing certain types of attacks or focusing on one layer of the OSI stack. In operation, one would desire to create a systematic security and authentication scheme whereby a claimed identity is vetted through physical-layer authentication.

References

Agrawal, D., Archambeault, B., Rao, J.R., & Rohatgi, P. (2003). The EM Side—Channel(s). *Cryptographic Hardware and Embedded Systems—CHES 2002, 2523,* 29–45.

Ahmed, I., Obermeier, S., Naedele, M., & Richard III, G. (2012). Scada systems: Challenges for forensic investigators. *Computer, 45*(12), 44–51.

Badenhop, C.W., Ramsey, B.W., Mullins, B.E., & Mailloux, L.O. (2016). Extraction and analysis of non-volatile memory of the ZW0301 module, a Z-Wave transceiver. *Digital Investigation, 17,* 14–27.

Bihl, T.J. (2015). *Feature Selection and Classifier Development for Radio Frequency Device Identification.* Ph.D. Dissertation, Air Force Institute of Technology, Wright-Patterson Air Force Base, OH.

Bihl, T.J., Bauer, K.W., & Temple, M.A. (2016). Feature selection for RF fingerprinting with multiple discriminant analysis and using ZigBee device emissions. *IEEE Transactions on Information Forensics and Security, 11*(8), 1862–1874.

Bihl, T.J., Young, W.A., & Weckman, G.R. (2016). Defining, understanding, and addressing big data. *International Journal of Business Analytics (IJBAN), 3*(2), 1–32.

Byres, E., & Lowe, J. (2004). The myths and facts behind cyber security risks for industrial control systems. *Proceedings of the VDE Kongres*, 213–218.

Candore, A., Kocabas, O., & Koushanfar, F. (2009). Robust stable radiometric fingerprinting for wireless devices. *IEEE International Workshop on Hardware-Oriented Security and Trust (HOST)*, 43–49.

Cao, H., Leung, V., Chow, C., & Chan, H. (2009). Enabling technologies for wireless body area networks: A survey and outlook. *IEEE Communications Magazine, 47*(12), 84–93.

Carbino, T.J., Temple, M.A., & Bihl, T.J. (2015). Ethernet card discrimination using unintentional cable emissions and constellation-based fingerprinting. *International Conference on Computing, Networking and Communications (ICNC)*, Garden Grove, CA, 369–373.

Cárdenas, A.A., Amin, S., & Sastry, S. (2008). Research challenges for the security of control systems. *HotSec.*

Chang, A.-M., Kannan, P.K., & Fellow, S. (2003). Preparing for wireless and mobile technologies in government. *E-government*, 345–393.

Chen, T., & Abu-Nimeh, S. (2011). Lessons from stuxnet. *Computer, 44*(1), 91–93.

Clarke, G.R., Reynders, D., & Wright, E. (2004). *Practical Modern SCADA Protocols: DNP3, 60870.5 and Related Systems.* Burlington, MA: Newnes.

Cobb, W.E. (2011). *Exploitation of Unintentional Information Leakage from Integrated Circuits.* PhD Dissertation, Air Force Institute of Technology, Wright-Patterson Air Force Base, OH.

Cobb, W.E., Garcia, E.W., Temple, M.A., Baldwin, R.O., & Kim, Y.C. (2010). Physical layer identification of embedded devices using RF-DNA fingerprinting. *Military Communications Conference (MILCOM)*, 2168–2173.

Cobb, W.E., Laspe, E.D., Baldwin, R.O., Temple, M.A., & Kim, Y.C. (2012). Intrinsic physical-layer authentication of integrated circuits. *IEEE Transactions on Information Forensics and Security, 7*(1), 14–24.

Couch, L.W. (1993). *Digital and Analog Communication Systems* (4th ed.). New York: MacMillan.

Creery, A., & Byres, E. (2005). Industrial cybersecurity for power system and SCADA networks. *Industry Applications Society 52nd Annual Petroleum and Chemical Industry Conference*, 303–309.

Daneels, A., & Salter, W. (1999). What is SCADA? *International Conference on Accelerator and Large Experimental Physics Control Systems*, 339–343.

DeJean, G., & Kirovski, D. (2007). RF-DNA: Radio-frequency certificates of authenticity. *Cryptographic Hardware and Embedded Systems (CHES)*, Springer, Berlin, 346–363.

Di, J., & Smith, S. (2007). A hardware threat modeling concept for trustable integrated circuits. *IEEE Region 5 Technical Conference*, 354–357.

Dolezilek, D., & Schweitzer, E.O. (2000). *SEL Communications and Integration White Paper.* Pullman, WA: Schweitzer Engineering Laboratories.

Ellis, K.J., & Serinken, N. (2001). Characteristics of radio transmitter fingerprints. *Radio Science, 36*(4), 585–597.

Ellison, R.J., Fisher, D.A., Linger, R.C., Lipson, H.F., & Longstaff, T. (1997). *Survivable Network Systems: An Emerging Discipline*. Pittsburgh, PA: Software Engineering Institute, Carnegie-Mellon University.

Fernandez, J., & Fernandez, A. (2005). SCADA systems: Vulnerabilities and remediation. *Journal of Computing Sciences in Colleges, 20*(4), 160–168.

Fovino, I., Carcano, A., Masera, M., & Trombetta, A. (2009). Design and implementation of a secure modbus protocol. *Critical Infrastructure Protection, 311* 83–96.

Frenzel, L. (2013). *What's the difference between IEEE 802.15.4 and ZigBee wireless?* Retrieved November 11, 2014, from Electronic Design: http://electronicdesign. com/what-s-difference-between/what-s-difference-between-ieee-802154-and-zigbee-wireless.

Galloway, B., & Hancke, G.P. (2013). Introduction to industrial control networks. *IEEE Communications Surveys and Tutorials, 15*(2), 860–880.

Gomez, J.A. (2005). *Survey of SCADA SYSTEMS and visualization of a real life process*. MS Thesis, Linköping University, Linköping, Sweden.

Goverment Accountability Office (GAO). (2008). *TVA Needs to Address Weaknesses in Control Systems and Networks, GAO-08–526*. Washington, DC: US Government.

Grau, D., Zeng, L., & Xiao, Y. (2012). Automatically tracking engineered components through shipping and receiving processes with passive identification technologies. *Automation in Construction, 28*, 36–44.

Guajardo, J., Kumar, S.S., Schrijen, G.J., & Tuyls, P. (2008). Brand and IP protection with physical unclonable functions. *IEEE International Symposium on Circuits and Systems (ISCAS)*, 3186–3189.

Guin, U., Huang, K., DiMase, D., Carulli, J.M., Tehranipoor, M., & Makris, Y. (2014). Counterfeit integrated circuits: A rising threat in the global semiconductor supply chain. *Proceedings of the IEEE, 102*(8), 1207–1228.

Güngör, V., Sahin, D., Kocak, T., Ergüt, S., Buccella, C., Cecati, C., & Hancke, G. (2011). Smart grid technologies: Communication technologies and standards. *IEEE Transactions on Industrial Informatics, 7*(4), 529–539.

Gutierrez, R.J., Bauer, K.W., Boehmke, B.C., Saie, C.M., & Bihl, T.J. (2017). Cyber anomaly detection: Using tabulated vectors and embedded analytics for efficient data mining. *Journal of Algorithms and Computational Technology*.

Harmer, P.K. (2013). *Development of a Learning from Signals Classifier for Cognitive Software Defined Radio Applications*. PhD Dissertation (DRAFT), Air Force Institute of Technology, Wright-Patterson Air Force Base, OH.

Harmon, R.R., Castro-Leon, E.G., & Bhide, S. (2015). Smart cities and the internet of things. *Portland International Conference on Management of Engineering and Technology (PICMET)*, 485–494.

Higgs, M. (2000). Electrical SCADA systems from the operators perspective. *International Conference on Human Interfaces in Control Rooms, Cockpits and Command Centres*, 458–461.

Holland, W.S., Young, W.A., & Weckman, G.R. (2011). Facility RFID localization system based on artificial neural networks. *International Journal of Industrial Engineering: Theory, Applications and Practice, 18*(1), 16–24.

Hu, X., Wang, B., & Ji, H. (2013). A wireless sensor network-based structural health monitoring system for highway bridges. *Computer-Aided Civil and Infrastructure Engineering, 28*(3), 193–209.

IEEE. (2004). *Overview and Guide to the IEEE 802 LMSC*. New York: Institute of Electrical and Electronics Engineers.

Jain, A.K., Ross, A., & Prabhakar, S. (2004). An introduction to biometric recognition. *IEEE Transactions on Circuits and Systems for Video Technology, 14*(1), 4–20.

Jang-Jaccard, J., & Nepal, S. (2014). A survey of emerging threats in cybersecurity. *Journal of Computer and System Sciences, 80*(5), 973–993.

Jiang, R., Lu, R., Wang, Y., Luo, J., Shen, C., & Shen, X.S. (2014). Energy-theft detection issues for advanced metering infrastructure in smart grid. *Tsinghua Science and Technology, 19*(2), 105–120.

Kang, D.-j., & Robles, R.J. (2009). Compartmentalization of protocols in SCADA communication. *International Journal of Advanced Science and Technology, 8*, 27–36.

Karri, R., Rajendran, J., Rosenfeld, K., & Tehranipoor, M. (2010). Trustworthy hardware: Identifying and classifying hardware trojans. *Computer, 43*(10), 39–46.

Khatib, A.-R., Dong, Z., Qiu, B., & Liu, Y. (2000). Thoughts on future Internet based power system information network architecture. *IEEE Power Engineering Society Summer Meeting*, 155–160.

Klein, R.W. (2009). *Application of dual-tree complex wavelet transforms to burst detection and RF fingerprint classification.* PhD Dissertation, Air Force Institute of Technology, Wright-Patterson Air Force Base, OH.

Landt, J. (2005). The history of RFID. *IEEE Potentials, 24*(4), 8–11.

Liao, Q., Luo, X.R., Gurung, A., & Shi, W. (2015). A holistic understanding of non-users' adoption of university campus wireless network: An empirical investigation. *Computers in Human Behavior, 48*, 220–229.

Lohweg, V., Hoffmann, J.L., Dörksen, H., Hildebrand, R., Gillich, E., Hofmann, J., & Schaed, J. (2013). Banknote authentication with mobile devices. *IS&T/SPIE Electronic Imaging, 8665*, 1–14.

Lopez, J., Temple, M.A., & Mullins, B.E. (2014). Exploitation of HART wired signal distinct native attribute (WS-DNA) features to verify field device identity and infer operating state. *International Conference on Critical Information Infrastructures Security*, 24–30.

Luiijf, E.A., & Klaver, M.H. (2004). Protecting a nation's critical infrastructure: The first steps. *IEEE International Conference on Systems, Man and Cybernetics*, 1185–1190.

Majzoobi, M., Koushanfar, F., & Potkonjak, M. (2009). Techniques for design and implementation of secure reconfigurable PUFs. *ACM Transactions on Reconfigurable Technology and Systems (TRETS), 2*(1), 1–33.

McMaster, H.R. (2003). *Crack in the Foundation: Defense Transformation and the Underlying Assumption of Dominant Knowledge in Future War.* Carlisle, PA: US Army War College.

Melaragno, A., Bandara, D., Wijesekera, D., & Michael, J. (2012). Securing the ZigBee protocol in the smart grid. *Computer, 45*(4), 92–94.

Miller, B., & Rowe, D.C. (2012). A survey of SCADA and critical infrastructure incidents. *1st Annual Conference on Research in Information Technology*, 51–56.

Mo, Y., Kim, T.H.-J., Brancik, K., Dickinson, D., Lee, H., Perrig, A., & Sinopoli, B. (2012). Cyber–physical security of a smart grid infrastructure. *Proceedings of the IEEE, 100*(1), 195–209.

Montrose, M.I. (2004). *EMC and the Printed Circuit Board: Design, Theory, and Layout Made Simple.* New York, John Wiley & Sons.

Moteff, J., & Parfomak, P. (2004). *Critical Infrastructure and Key Assets: Definition and Identification.* Washington, DC: Congressional Research Service.

Nawir, M., Amir, A., Yaakob, N., & Lynn, O. (2016). Internet of Things (IoT): Taxonomy of security attacks. *3rd International Conference on Electronic Design (ICED)*, 321–326.

Ozdemir, E., & Karacor, M. (2006). Mobile phone based SCADA for industrial automation. *ISA transactions, 45*(1), 67–75.

Patton, M., Gross, E., Chinn, R., Forbis, S., Walker, L., & Chen, H. (2014). Uninvited connections: A study of vulnerable devices on the internet of things (IoT). *IEEE Joint Intelligence and Security Informatics Conference (JISIC)*, 232–235.

Peltier, T.R. (2005). *Information Security Risk Analysis*. New York: CRC Press.

PFP Cybersecurity. (2016). Embedding Security in the Internet of Things. *White Paper*, PFP Cyber Security, Vienna, VA.

Queiroz, C., Mahmood, A., Hu, J., Tari, Z., & Yu, X. (2009). Building a SCADA security testbed. *Third International Conference on Network and System Security*, 357–364.

Ramsey, B., Temple, M., & Mullins, B. (2012). PHY foundation for multifactor ZigBee node authentication. *Global Communication Conference (GLOBECOM)*, 795–800.

Rescorla, E. (2005). Is finding security holes a good idea? *IEEE Security & Privacy, 3*(1), 14–19.

Roberts, C.M. (2006). Radio frequency identification (RFID). *Computers & Security, 25*(1), 18–26.

Robles, R.J., & Choi, M.-k. (2009). Assessment of the vulnerabilities of SCADA, control systems and critical infrastructure systems. *International Journal of Grid and Distributed Computing Assessment, 2*(2), 27–34.

Robles, R.J., Choi, M.-k., & Kim, T.-h. (2009). The taxonomy of SCADA communication protocols. *Proceedings of KIIT Summer Conference*, 116–119.

Ross, B.P., Carbino, T.J., & Stone, S.J. (2017). Physical-layer discrimination of power line communications. *International Conference on Computing, Networking and Communications (ICNC)*, 341–345.

Ryer, D.M., Bihl, T.J., Bauer, K.W., & Rogers, S.K. (2012). QUEST hierarchy for hyperspectral face recognition. *Advances in Artificial Intelligence* 2012, 13.

Samuelson, D.A. (2016). Using big data in cybersecurity. *ORMS-Today, 43*(5).

Sandaruwan, G.P., Ranaweera, P.S., & Oleshchuk, V.A. (2013). PLC security and critical infrastructure protection. *8th IEEE International Conference on Industrial and Information Systems (ICIIS)*, 81–85.

Scanlon, P., Kennedy, I.O., & Liu, Y. (2010). Feature extraction approaches to RF fingerprinting for device identification in femtocells. *Bell Labs Technical Journal, 15*(3), 141–151.

Schneider Electric. (2012). *SCADA Systems*. Rueil-Malmaison, France: Schneider Electric.

Slay, J., & Miller, M. (2007). Lessons learned from the Maroochy water breach. *Critical Infrastructure Protection*, 253 73–82.

Smith, R. (2014). Assault on California power station raises alarm on potential for terrorism. *Wall Street Journal*.

Snow, A.P., Varshney, U., & Malloy, A.D. (2000). Reliability and survivability of wireless and mobile networks. *Computer, 33*(7), 49–55.

Stuttard, D. (2005). Security & obscurity. *Network Security, 7*, 10–12.

Suski, W.C., Temple, M.A., Mendenhall, M.J., & Mills, R.F. (2008). Using spectral fingerprinting to improve wireless network security. *IEEE Global Communications Conference (GLOBECOM)*, 1–5.

Tehranipoor, M.M., Guin, U. & Forte, D. (2015). *Counterfeit Integrated Circuits*, (pp. 15–36). Springer International Publishing.

Thirukkonda, S. (2009). *Correlation in Firm Default Behavior*. MS Thesis, Massachusetts Institute of Technology.

Tomko, A.A., Rieser, C.J., & Buell, L.H. (2006). Physical-layer intrusion detection in wireless networks. *IEEE Military Communications Conference (MILCOM)*, 1–7.

Ultes-Nitsche, U., & Yoo, I. (2004). Run-time protocol conformance verification in firewalls. *ISSA*, 1–11.

US Government Accountability Office. (2004). *Critical Infrastructure Protection Challenges and Efforts to Secure Control Systems*. Washington, DC: GAO–05–434.

Walton, R., & Limited, W.-M. (2006). Balancing the insider and outsider threat. *Computer Fraud & Security, 11*, 8–11.

Weber, M., Birkel, U., Collmann, R., & Engelbrecht, J. (2010). Comparison of various methods for indoor RF fingerprinting using leaky feeder cable. *Workshop on Positioning Navigation and Communication (WPNC)*, 291–298.

Weng, H., Dong, X., Hu, X., Beetner, D.G., Hubing, T., & Wunsch, D. (2005). Neural network detection and identification of electronic devices based on their unintended emissions. *International Symposium on Electromagnetic Compatibility, 1*, 245–249.

Wortmann, F., & Flüchter, K. (2015). Internet of things. *Business & Information Systems Engineering, 57*(3), 221–224.

Xing, K., Srinivasan, S., Jose, M., Li, J., & Cheng, X. (2010). Attacks and countermeasures in sensor networks: A survey. *Network Security*, 251–272.

Yang, H., Luo, H., Ye, F., Lu, S., & Zhang, L. (2004). Security in mobile ad hoc networks: Challenges and solutions. *IEEE Wireless Communications, 11*(1), 38–47.

Zhu, B., & Sastry, S. (2010). SCADA-specific intrusion detection/prevention systems: A survey and taxonomy. *Workshop on Secure Control Systems (SCS)*.

6

Data-Mining Methods for Electricity Theft Detection

Trevor J. Bihl

Wright State University

Ahmed F. Zobaa

Brunel University London

CONTENTS

6.1 Introduction 107
6.2 Transmission and Distribution System Losses 108
6.3 Electricity Theft Methods 110
    6.3.1 Fraud 111
        6.3.1.1 Bypassing Existing Meetings 111
        6.3.1.2 Meter Tampering 111
    6.3.2 Billing Issues 114
    6.3.3 Outright Theft 115
    6.3.4 Electricity Theft and Data Collection 116
6.4 Data Mining and Electricity Theft 116
    6.4.1 Prediction 117
    6.4.2 Classification and Clustering 117
    6.4.3 Detection 118
6.5 Issues and Directions in Electricity Theft-Related Data-Mining
    Research 118
6.6 Conclusions 120
Bibliography 120

6.1 Introduction

Electricity theft involves the intentional theft, or nonpayment, of electrical services. Worldwide electricity theft losses are significant and estimated recently (2014) at $86.3 billion a year (Northeast Group, 2014). Detecting and mitigating electricity theft has historically involved a combination of usage analysis (Preece, 1882; Goldman & Sweet, 2008), improving the physical security of meters (Haskins, 1897; Nesbit, 2000), and inspection of meters and premises (Nesbit, 2000). However, the advent of smart meters and data mining has enabled utilities to employ sophisticated methods to find potential theft (Depuru, Wang, & Devabhaktuni, 2011a,b). Of interests herein is developing an understanding of what data-mining methods have been applied for electricity theft detection and their general performance results. In applying data mining to electricity theft, one generally attempts to find losses that do not conform to expected losses. Thus, one needs to have a firm understanding of losses to understand electricity theft; for this, knowledge of both losses in electrical transmission and distribution (T&D) systems and of electricity theft methods is needed. The authors conclude by discussing limitations in current data mining for electricity theft research and aim to provide an understanding of opportunities for data mining and data science in electricity theft detection.

6.2 Transmission and Distribution System Losses

T&D systems include high-voltage and long-distance transmission lines and components which link generation to the distribution system; the distribution system is associated with relatively lower voltage components which distribute electricity in a relatively small area (Heydt, 2010). Due to physical properties of devices and components, losses are inherent in T&D systems and are the difference between total produced kilowatt hours and total billed/consumed kilowatt hours.

T&D system losses can be divided into two groups: *technical losses* and *nontechnical losses* (Davidson, Odubiyi, Kachienga, & Manhire, 2002; Suriyamongkol, 2002; Dortolina & Nadira, 2005). Technical losses are due to physical properties of the T&D system, i.e., resistances and inefficiencies, while nontechnical losses (NTLs) are due to nonphysical means, e.g., electricity theft, accounting errors, and faulty readings (Bihl & Hajjar, 2017). Examples of both types of losses are found in Table 6.1, as compiled by Bihl and Hajjar (2017).

While electricity theft is of primary importance, it is naturally hard to separate from the other NTLs and thus in general one is interested estimating in NTLs. Herein, detecting electricity theft and detection NTLs are essentially synonymous. However, one must exercise some care when NTLs are detected because not all NTLs are due to electricity theft.

Appropriate estimates of NTLs can logically impact prior probability estimates for data mining and provide a rough understanding of the severity of the issue. To estimate NTLs, it is best to begin by estimating technical losses, which are easier to constrain since they are bound by physical properties of T&D systems (Davidson, Odubiyi, Kachienga, & Manhire, 2002). Even so, the technical losses in T&D systems will vary by country, region, utility, and

TABLE 6.1

Examples of Technical and Nontechnical Losses

Technical Losses		Nontechnical Losses
Variable	Fixed	
Load	Hysteresis	Accounting errors
Series	Core	Electricity theft
Copper	Eddy current	Faulty meters (inaccurate and miscalibrated)
Transport related	No load	Faulty meter-reading methods
	Shunt	Incorrect meter readings
	Iron	Technical loss computation errors

Source: From Bihl and Hajjar (2017).

estimate. As an example, relative to the United States, rough estimates on T&D losses include the following: 6%–9% (Dortolina & Nadira, 2005), 7.6% (Gustafson & Baylor, 1989), 8% (Farhangi, 2010), and 10% (Weslowski, 1976).

Since NTL quantities are general unknown, it is inherently difficult to estimate their quantity since approximations and arbitrary estimates exist in the "known" technical losses (Davidson, Odubiyi, Kachienga, & Manhire, 2002). Thus, Nesbit (2000) reported that there is no known true percentage loss, with various regional, technical, and cultural differences driving disparate theft rates across a country. However, it is possible to determine rough upper and lower bounds for theft using electric generation and consumption data.

At the macro level, the total T&D loss (in kilowatt-hours, or any other appropriate units) can be thought of as

$$W_{\text{NetGen}} = W_{\text{Sold}} + W_{\text{T\&D losses}} \tag{6.1}$$

where W_{NetGen} are net kilowatt-hours generated, W_{Sold} is the total amount of electricity sold, and $W_{\text{T\&D losses}}$ are the kilowatt-hours lost to various T&D losses. The percentage of T&D losses as a percentage of generation can be estimated as follows:

$$\%_{\text{T\&D losses}} = \frac{(W_{\text{NetGen}} + W_{\text{Sold}})}{W_{\text{NetGen}}} \times 100 \tag{6.2}$$

where $\%_{\text{T\&D losses}}$ are the total percentage of T&D losses (Donziger, 1979). To calculate NTLs (in the equations), one must make an assumption that $W_{\text{T\&D losses}} = W_{\text{T\&D TL}} + W_{\text{T\&D NTL}}$, and thus we have

$$W_{\text{NetGen}} = W_{\text{Sold}} + W_{\text{T\&D TL}} + W_{\text{T\&D NTL}} \tag{6.3}$$

which is consistent with Doorduin, Mouton, Herman, and Beukes (2004), which further means that one can solve for $\%_{\text{T\&D NTL}}$,

$$\%_{T\&D\,NTL} = \%_{T\&D\,losses} - \%_{T\&D\,TL} \tag{6.4}$$

where $\%_{T\&D\,NTL}$ and $\%_{T\&D\,TL}$ are the respective percentages of NTLs and TL. The formulation in Equations (6.1)–(6.4) is consistent with that of Donziger's (1979) first known economic analysis of theft in utility sectors (gas, electric, telephone, etc.) for the USA. This T&D calculation is also functionally identical to ones independently derived by others (Davidson, Odubiyi, Kachienga, & Manhire, 2002; Davidson, 2003; Anderson, 2006).

To apply Equation (6.4), one must have a reasonable estimate of losses and losses associated with technical losses. Estimating $\%_{T\&D\,losses}$ should be relatively trivial since all quantities in Equation (6.2) are apparent from records and estimating technical losses is the primary challenge in determining NTLs. Donziger (1979), focusing on the United States, calculated $\%_{T\&D\,losses}$ as 6.35% and assumed that $\%_{T\&D\,TL}$ was 7%, yielding an estimate of $\%_{T\&D\,NTL}$ of 2.35%.

6.3 Electricity Theft Methods

To understand how to detect electricity theft, one must understand the various methods employed. As discussed in Bihl and Hajjar (2017), and consistent with Hale (1896), Wilson (1988), Nesbit (2000), Smith (2004), and Dey et al. (2010), electricity theft is of three types:

A. *Outright Theft*, which can be divided by:
  - *Tapping* an overhead line to create a new, illegal connection
  - *Induction Coupling* whereby energy from a power line is collected by electromagnetic induction without physically connecting to the line.

B. *Fraud*, which is accomplished by
  - *Bypassing* a meter to prevent it from measuring the power consumed
  - *Tampering* with a meter to cause it to output a more favorable reading for the customer. This is subdivided into mechanical and digital/smart meter methods

C. *Billing Issues*
  - Deliberate nonpayment of bills.
  - *Billing irregularities*, both intentional (bribing officials to ignore use) and unintentional (accounting errors and faulty meters, faulty meter-reading methods, incorrect meter readings, technical loss computation errors) which account for most other NTLs

Although only one method is termed as explicitly theft, all of these issues involve consuming electricity which is not paid for. As will become apparent through discussing these issues, various challenges and opportunities exist for detecting theft across these methods.

6.3.1 Fraud

Electricity theft methods that aim to reduce the recorded consumption of electricity are viewed as fraud (Smith, 2004; Bihl & Hajjar, 2017). Here, an electricity thief is either a past or present customer of a utility and aims to bypass or tamper with metering equipment to reduce their bill. During whatever process a thief might use to commit fraud, significant risk is also taken by the thief since it is likely that all illegal connections and wiring was performed using live wires (it is not likely that a thief would notify a utility to request a power shutoff) (Bihl & Hajjar, 2017). This risk can logically result in injury, death, or damages to the thief and the premises; thus, electricity theft introduces significant risks beyond the financial.

6.3.1.1 Bypassing Existing Meetings

Bypassing a meter involves creating a direct connection for a premise to the electric grid, whereby a connection is made around the meter. Figure 6.1 provides two examples of bypassing, through the use of automotive jumper cables, Figure 6.1a, and screwdrivers, Figure 6.1b. Approaches to bypassing a meter include completely disconnect the meter and placing wires or metal in the meter connections (US Patent No. 2,019,866, 1933), as seen in Figure 6.1a. Alternatively, one can leave the meter connected in addition to the bypass whereby the meter records less usage than actual since the meter has more resistances than the bypass connections (Hallberg, 1905a,b; Seger & Icover, 1988), as seen in Figure 6.1b where screwdriver were placed behind the meter to provide a direct connection to the line. Figure 6.2 provides an illustration of the operation of bypassing; as seen in this figure, the bypass line does not include a meter and thus it has less-resistance resulting in a meter recording less usage (Hallberg, 1905a,b; Weslowski, 1976; Wilson, 1988) (US Patent No. 2,019,866, 1933). Approaches to prevent bypassing include reducing access to the lines (Wade, 1955).

6.3.1.2 Meter Tampering

An additional vector to electricity theft exists whereby thieves modify or damage electric meters to read less usage. Both mechanical and electronic, i.e., Smart Meters, are susceptible to tampering by electricity thieves and tampering has been a concern for over 100 years (cf. Haskins, 1897; Nesbit, 2000). Preventing meter tampering involves using seals to prevent and indicate unauthorized access (US Patent No. 1,612,420, 1926), providing robust

(a)

(b)

FIGURE 6.1
Examples of bypassing a metering through the use of automobile jumper cables (a) and screw-drivers (b).

security around meters (Clark, 1928), and including sensors to detect access (US Patent No. 4,565,995, 1986).

Mechanic meters have been in use for over 100 years and methods of theft have changed little in that time, see (Bihl & Hajjar, 2017). Figure 6.3 presents a conceptual mechanical meter with locations of theft highlighted, per Suriyamongkol (2002). In general, options for tampering with the meter include, as discussed in Wilson (1988), Nesbit (2000), Suriyamongkol (2002), and Bihl and Hajjar (2017), limiting disc movement, tampering with the calibration, using magnets to disrupt operation, contaminating the enclosure and parts, disconnecting the neutral conductor, and damaging the movement.

Beyond susceptibilities to theft, various limitations exist with mechanical meters, such as their inability to store usage data. Digital meters, considered herein as any electronic or smart meter, have been presented as an improved solution over mechanical meters. Digital meters generally operate consistent

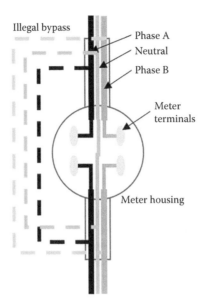

FIGURE 6.2
Example of meter bypass showing typical two-phase connection, i.e., United States houses, with a bypass making a connection around the meter.

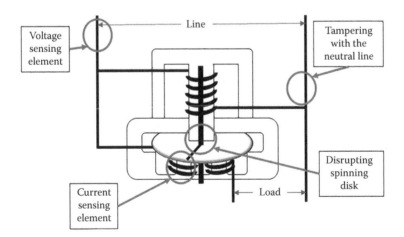

FIGURE 6.3
Conceptualization of a single-phase mechanical watt-hour meter with locations susceptible to theft highlighted. (Adapted from Suriyamongkol, 2002.)

with the conceptualization presented in Figure 6.4; here current and voltage sensors monitor the power flow into a premise. Analog-to-digital (ADC) converters are used to convert continuously variable sensor readings to discrete values which are analyzed by digital signal processor (DSP) (Sreenivasan, 2011). The DSP is programmed with an algorithm that measures energy

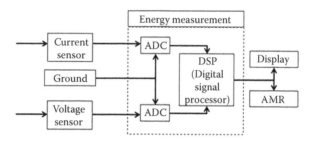

FIGURE 6.4
Conceptualization of a digital meter. (Adapted from Sreenivasan, 2011.)

usage and the resultant usage is displayed visually (Sreenivasan, 2011). An automatic meter reading (AMR) is also likely included and used to communication with the utility for remote meter-reading purposes (Sreenivasan, 2011). Since they can be monitored remotely, digital meters also have potential manpower and cost reduction benefits by reducing physical meter readings (Tan, Lee, & Mok, 2007).

Although digital meters have been proposed as solutions to electricity theft (Cavdar, 2004), they are in fact susceptible to electricity theft (McDaniel & McLaughlin, 2009; McLaughlin, Holbert, Fawaz, Berthier, & Zonouz, 2013). While digital meters avoid many vulnerabilities associated with mechanical meters since there are no moving parts (Singhal, 1999), they introduce new vulnerabilities and further ethical issues such as consumer privacy (McDaniel & McLaughlin, 2009). Vectors for electricity theft which are introduced by digital meters include: flooding the bandwidth and exhausting member with cyber-attacks, modifying firmware on the meters, stealing credentials or physically extracting passwords, intercepting and altering communication, compromising through remote network exploits or attacking optical ports, stopping data logging, and altering data logs (McLaughlin, Holbert, Fawaz, Berthier, & Zonouz, 2013).

Digital meters are largely considered to be very insecure cyber technology (Cimpanu, 2017), as they introduce vectors for hacking (Goel & Hong, 2015), and include vulnerabilities to data injection attacks that can change the recorded usage data (Wang, Liang, Mu, Wang, & Zhang, 2015). Additionally, some digital meters are susceptible to physical tampering whereby a disconnected ground wire can stop some meters from recording usage (Dey et al., 2010). However, the overall advantages of digital meters are viewed as overcoming their limitations (Depuru, Wang, & Devabhaktuni, 2011a,b).

6.3.2 Billing Issues

NTLs and theft can also be present through billing issues. These manifest as either irregularities or deliberate nonpayment by customers (Smith, 2004). Irregularities generally occur due to poor accounting practices or bribery

(Smith, 2004). Both of which are logically alleviated via digital meters, remote meter reading, and improved accountability (Appleyard, 1963; Ghajar & Khalife, 2003). In contrast to irregularities, deliberate nonpayment involves active theft whereby a customer refuses to pay an electric bill. While deliberate nonpayment inefficient as a method of theft since records would exist of the nonpaying customers, it is a viable form of theft since utilities cannot always collect unpaid bills. Additionally, it is not always cost effective to pursue recovering losses since there is a causal relationship between this method of theft and economic problems of customers (Smith, 2004).

6.3.3 Outright Theft

One final form of electricity theft exists whereby electricity is stolen directly either by unauthorized direct connections or by indirect, e.g., induction, connections to the T&D grid. In both cases, electricity theft involves creating a new and illegal connection directly to a T&D system without the approval of the utility. Thus, one could connect to the power grid without being a past or present customer of a utility.

Direct connections can be made by "tapping" an overhead line. An example is presented in Figure 6.5, where "tapped" is employed by creating a

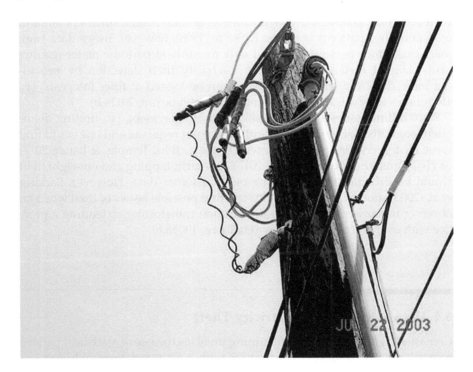

FIGURE 6.5
Contemporary example of tapping in America.

direct connection to a power line by a jumper cable directly connects house wiring to overhead lines. Tapping has a long history of being a serviceable electricity theft method (cf. Hallberg, 1905a,b; Wilson, 1988) and involves a risk of electrocution to the thief and bystanders and damages to T&D system components (Kim, 2011).

Theft by induction involves stealing electricity via induction coupling. It is considered generally improbable to individually steal a significant amount of electricity due to the large investment required in copper need to create a sufficiently large coil (Deardorff, 2006). However, recent developments by Siegel (2017) have aimed at creating small "free electricity" devices which steal energy through induction from power lines for small electronic device charging. Widespread adoption of this method of electricity theft by the masses, as advocated by many (Dansie, 2013; von der Gracht, Salcher, & Graf Kerssenbrock, 2016), could result in significant electricity theft and reduced abilities of utilities to plan.

6.3.4 Electricity Theft and Data Collection

To apply data mining to detect electricity theft, one logically needs to collect appropriate data for analysis. While digital meters are not immune to theft, they have distinct advantages over mechanical meters for electricity theft detection. Data-mining algorithms in general need a sufficient amount of data to find patterns (e.g., Lee & Stolfo, 1998); however, usage data from mechanical meters is often logged only monthly at periodic meter-reading visits. Digital meters can facilitate electricity theft detection by providing more data to analyze since usage can be logged at finer intervals, i.e., 15-minutes intervals (Depuru, Wang, & Devabhaktuni, 2011a,b).

Since billing issues are largely related to nonpayment, i.e., finding delinquent accounts, data-mining algorithms are not requisite and one could find these customers via spreadsheet searches (e.g., Bihl, Temple, & Bauer, 2017) or visual analytics (e.g., Koitzsch, 2017). Similarly, tapping and out-right theft would be difficult to detect using only customer data. However, Bandim et al. (2003) showed how algorithms can find possible issues of theft when an observer meter is placed near a distribution transformer, extending a practice with over 100 years of history (Hallberg, 1905a,b).

6.4 Data Mining and Electricity Theft

Consistent with Chapter 2, data mining involves the use of statistical pattern recognition algorithms to find meaning within a given dataset. While statistical analysis tends to focus on primary data analysis, e.g., collecting data to answer a specific question, data mining is a form of secondary data analysis, whereby data are broadly collected and one later attempts to find meaning-ful patterns (Hand, 1998; Bihl, Young, & Weckman, 2016). Data mining is incidentally synonymous with the following approaches (Bihl, Young, & Weckman, 2016): knowledge discovery in databases (KDD) (Mannila, 1996), data dredging, knowledge extraction, knowledge mining, data archaeology (Chen, Han, & Yu, 1996), and fishing (Hand, 1998). Within data mining, vari-ous algorithmic approaches can be used to find meaning. At a high level one is interested in many different problems in data mining, including (Bose & Mahapatra, 2001; Bihl, Young, & Weckman, 2016): *dimensionality reduction, visualizations, prediction, classification, association,* and *detection.* For electricity theft detection research, primary methods of interest include the *prediction* of usage patterns, *detection* of anomalous usage, *associating* (clustering) usage patterns together, and *classifying* customers as thieves.

6.4.1 Prediction

Prediction involves developing a mathematical model from data which can be used to predict patterns or values. When considering electricity usage data, prediction algorithms can be used for load forecasting to facilitate planning (Hagan & Behr, 1987; Ghofrani, Hassanzadeh, Etezadi-Amoli, & Fadali, 2011). Prediction algorithms can take the form of various curve fitting methods, such as simple linear regression and polynomial models (Amral, Ozveren, & King, 2007), to nonlinear and complex algorithms, e.g., nonlinear autoregressive neural networks (Ward, Bihl, & Bauer, 2014).

For electricity theft detection using prediction models, a variety of predic-tion algorithms have been proposed, including Auto-Regressive Integrated Moving Average (ARIMA) (Krishna, Iyer, & Sanders, 2015), state estima-tion and load profiling (Viegas, Esteves, Melício, Mendes, & Vieira, 2017), and regression (Monedero et al., 2010; Yip, Tan, Tan, Gan, & Bakar, 2017). For electricity theft detection, prediction algorithms can be used to predict expected consumption for honest consumers and flag potential theft cus-tomer records which deviate significantly from expected, i.e., when observed usage falls outside a confidence interval (Krishna, Iyer, & Sanders, 2015), regression coefficients deviate strongly from expected (Yip, Tan, Tan, Gan, & Bakar, 2017), and correlation coefficients associate with continuous declines in usage (Monedero et al., 2010).

6.4.2 Classification and Clustering

Classification involves two general types of approaches, either supervised learning, where examples of groups are known, or unsupervised learning, where known groupings in the data are not known (Jain, Duin, & Mao, 2000). Supervised methods consider labeled data, e.g., normal usage and abnormal usage groups, to learn patterns within the data that create decision bound-aries that separate groups (Jain, Duin, & Mao, 2000). To analyze a dataset to find previously unknown patterns, e.g., types of electricity usage patterns, one is generally interested in clustering algorithms which find regions of similar data samples (Jain, Duin, & Mao, 2000). In both unsupervised and supervised methods, once an algorithm is trained, future observations are then processed and assigned to the best matching group.

Of interest in classification is determining membership of observations into an *electricity theft* group; however, such a categorization implies the existence of two possible subjective states: the *electricity theft* group and the *non-theft* group of honest consumers (Cox, 1979; Fry, 2012). For electricity theft detection, classification algorithms have largely focused on support vector machines (SVMs), artificial neural networks (ANNs), decision trees (DTs), rule-based systems (RBS), optimum-path forest, naïve Bayes, and *K*-means clustering (Viegas, Esteves, Melício, Mendes, & Vieira, 2017).

6.4.3 Detection

Detection involves finding a signal of interest in the presence of noise (Kelly, 1986). One variant of detection is anomaly detection which involves finding statistical anomalies, samples that are considerably different from the majority of samples (Duda, Hart, & Stork, 2012). For electricity theft detection, anomaly detection methods can be used to find load profiles which are statistically different from normal, and majority, profiles (McLaughlin, Holbert, Fawaz, Berthier, & Zonouz, 2013).

6.5 Issues and Directions in Electricity Theft-
   Related Data-Mining Research

Although a wide variety of data-mining methods exist in the literature (cf. Jain, Duin, & Mao, 2000; Wu et al., 2008; Duda, Hart, & Stork, 2012), the application of data mining to electricity theft detection has focused on only a narrow scope of the available methods. As discussed by Viegas, Esteves, Melício, Mendes, and Vieira (2017) and consistent with Jiang et al. (2014), while 91 data mining or statistical methods have been used in electricity theft detection research the domain is more focused. Largely, 14 methods have seen consistent and repeated use: SVMs, ANNs, DTs, RBS, optimum-path forest, naïve Bayes, *K*-means clustering, load profiling, direct calculation, state estimation, technical loss modeling, feature section, and text mining (Viegas, Esteves, Melício, Mendes, & Vieira, 2017).

Repeatability is a concern in data-mining research (e.g., Zhang, 2007), and irreproducibility issues make comparisons of performance and claims difficult. Various reproducibility issues exist in contemporary published electricity theft data-mining research and can be grouped into a few varieties.

In analyzing the literature, the authors have identified six general issues that exist in electricity theft data-mining research, as summarized in Table 6.2. Example references are included, but it should be noted that reproducibility issues are very common in contemporary electricity theft data-mining research.

As presented in Table 6.2, another factor limiting reproducibility is that performance measures are either not always included or are not consistent across publications. This was evident in the comparisons of Glauner et al. (2016) and Jiang et al. (2014), where standard classification accuracy measures (see Fawcett, 2006) are not available for many studies. The result of this is that comparisons are difficult across studies.

Further issues exist due to the "black box" and automated nature of many data-mining methods, whereby algorithmic setting can be opaque to both users and readers. As discussed in Zhang (2007), relative to ANNs, the result of these issues is that, even if one had the same data and algorithm, it might still be impossible to reproduce results since the appropriate algorithmic settings are unknown.

The proprietary nature of real electricity usage data introduces additional issues. When proprietary data are used, researchers have less ability to verify claims or improve over published results. Additionally, the lack of availability of real-world electricity theft data pushes some researchers to creating simulated data. However, simulated data may not be consistent with real-world usage data. It would be beneficial if researchers had access to usage data that is collected directly by utilities. The availability of such a dataset

TABLE 6.2

Issues in Data-Mining Method Exploitation for Electricity Theft Detection

Issues	General Description	Example Reference
No performance measures discernible	Data mining used, but performance not reported	Cabral, Pinto, and Pinto (2009)
Unclear performance measures	Accuracy reported, but type of accuracy is unclear	Depuru, Wang, and Devabhaktuni (2011a,b)
Algorithmic details missing	Data mining used, but experimental details for algorithm settings not reported	Nagi, Mohammad, Yap, Tiong, and Ahmed (2008)
Proprietary data	Repeatability becomes extremely difficult since only the authors have access to the data (not always explicitly stated)	Cabral, Pinto, and Pinto (2009)
Insufficient data details	Characteristics of the dataset (features, distribution of groups, source) are not sufficiently reported	Nizar, Dong, and Wang (2008)
Fabricated data	Contrived data that might not reasonably capture real-world electricity theft characteristics	Suriyamongkol (2002)

would alleviates many such issues and enable comparisons with similar studies. Additionally, such research further makes any related research results transitionable and realistic.

6.6 Conclusions

Electricity theft is a major concern for utilities in planning and revenue protection. The advent of digital metering has enabled the exploitation of data-mining methods for electricity theft detection. Current work has focused on a variety of prediction, classification, and detection methods. While this research domain has seen many methods applied, the repeatability of methods and availability of data are issues hampering research. Thus, the authors also advocate further improvements in the rigorous application of the wealth of methods and best practices available in data mining to this area.

Bibliography

Amral, N., Ozveren, C., & King, D. (2007). Short term load forecasting using multiple linear regression. *42nd International Universities Power Engineering Conference (UPEC)*, Brighton, UK, 1192–1198.

Anderson, M. (2006). *How to Identify Electricity Theft in Apartments without Hardware or Software Investments*. BluTrend LLC, Atlanta, Georgia.

Appleyard, V.A. (1963). Remote reading of meters. *Journal (American Water Works Association)*, 55(10), 1289–1291.

Bandim, C., Alves, J., Pinto, A., Souza, F., Loureiro, M., Magalhaes, C., & Galvez, D. (2003). Identification of energy theft and tampered meters using a central observer meter: A mathematical approach. *Proceedings of the IEEE PES Transmission and Distribution Conference and Exposition*, Dallas, TX, 163–168.

Bihl, T.J., & Hajjar, S. (2017). Electricity theft concerns within advanced energy technologies. *IEEE National Aerospace & Electronics Conference (NAECON)*, Dayton, OH.

Bihl, T.J., Young II, W.A., & Weckman, G.R. (2016). Defining, understanding, and addressing big data. *International Journal of Business Analytics (IJBAN)*, 3(2), 1–32.

Bihl, T., Temple, M., & Bauer, K. (2017). An optimization framework for generalized relevance learning vector quantization with application to Z-wave device fingerprinting. *Hawaii International Conference on System Sciences*, Waikoloa, HI.

Bose, I., & Mahapatra, R. (2001). Business data mining—A machine learning perspective. *Information & Management*, 39(3), 211–225.

Cabral, J.E., Pinto, J.O., & Pinto, A.M. (2009). Fraud detection system for high and low voltage electricity consumers based on data mining. *Power & Energy Society General Meeting*, 1–5.

Cavdar, I.H. (2004). A solution to remote detection of illegal electricity usage via power line communications. *IEEE Power Engineering Society General Meeting*, 896–900.

Chen, M.-S., Han, J., & Yu, P.S. (1996). Data mining: An overview from a database perspective. *IEEE Transactions on Knowledge and Data Engineering, 8*(6), 866–883.

Cimpanu, C. (2017, January 5). *Smart meters are laughably insecure, are a real danger to smart homes.* Retrieved from Bleeping Computer: www.bleepingcomputer.com/news/security/smart-meters-are-laughably-insecure-are-a-real-danger-to-smart-homes/.

Clark, S.B. (1928). Iron-clad services protect against theft. *Electrical World, 91*(7), 347.

Cox, R.T. (1979). Of inference and inquiry, an essay in inductive logic. *Proceedings of the Maximum Entropy Formalism. MIT Press*, Cambridge, MA, 119–168.

Dansie, M. (2013, June 28). Free Electricity From Thin Air. Retrieved June 10, 2017, from Revolution Green: http://revolution-green.com/free-electricity-from-thin-air/.

Davidson, I.E. (2002, October). Evaluation and effective management of nontechnical losses in electrical power networks, In *Africon Conference in Africa, 2002. IEEE AFRICON. 6th*, George, South Africa, Vol. 1, pp. 473–477.

Davidson, I.E., Odubiyi, A., Kachienga, M.O., & Manhire, B. (2002, April). Technical loss computation and economic dispatch model for T&D systems in a deregulated ESI. *Power Engineering Journal, 16*(2), 55–60.

Davis, W.S. (1926, December 28). Means for precluding tampering with electric meters, *US Patent No. 1,612,420*.

Deardorff, D.L. (2006, Summer). A Solution to the RWP for Exam 1—Stealing Power. Retrieved August 28, 2015, from Physics 25: http://user.physics.unc.edu/~deardorf/phys25/rwp/exam1rwpsolution.html.

Depuru, S.S., Wang, L., & Devabhaktuni, V. (2011a). Smart meters for power grid: Challenges, issues, advantages and status. *Renewable and Sustainable Energy Reviews, 15*(6), 2736–2742.

Depuru, S.S., Wang, L., & Devabhaktuni, V. (2011b). Support vector machine based data classification for detection of electricity theft. *IEEE/PES Power Systems Conference and Exposition (PSCE)*, 1–8.

Dey, H.S., ul-Mamun, M., Shahadat, M., Ahamed, A., Ahamed, S.U., & Arefin, K.S. (2010). Design and implementation of a novel protection device to prevent tampering and electricity theft in commercial energy meters. *Journal of Computer and Information Technology, 1*(1), 88–94.

Donziger, A.J. (1979, September 22). The underground economy and the theft of utility services. *Public Utilities Fortnightly*, 23–27.

Doorduin, W.A., Mouton, H.T., Herman, R., & Beukes, H.J. (2004). Feasibility study of electricity theft detection mobile remote check meters. *AFRICON Confernece in Africa, 1*, 373–376.

Dortolina, C.A., & Nadira, R. (2005). The loss that is unknown is no loss at all: A top-down/bottom-up approach for estimating distribution losses. *IEEE Transactions on Power Systems, 20*(2), 1119–1125.

Duda, R., Hart, P., & Stork, D. (2012). *Pattern Classification*. John Wiley & Sons, New York.

Farhangi, H. (2010). The path of the smart grid. *IEEE Power and Energy Magazine, 8*(1), 18–28.

Fawcett, T. (2006). An introduction to ROC analysis. *Pattern Recognition Letters, 27*(8), 861–874.

Fry, R. (2012, January 26). *Qualia, Intelligence, and Computation.* Air Force Institute of Technology (AFIT) Guest Lecture.

Ghajar, R.F., & Khalife, J. (2003). Cost/benefit analysis of an AMR system to reduce electricity theft and maximize revenues for Electricite du Liban. *Applied Energy, 76*(1), 25–37.

Ghofrani, M., Hassanzadeh, M., Etezadi-Amoli, M., & Fadali, M.S. (2011). Smart meter based short-term load forecasting for residential customers. *North American Power Symposium (NAPS)*, 1–5.

Glauner, P., Boechat, A., Dolberg, L., Meira, J., State, R., Bettinger, F., & Duarte, D. (2016). The challenge of non-technical loss detection using artificial intelligence: A survey. arXiv preprint arXiv:1606.00626.

Goel, S., & Hong, Y. (2015). Security challenges in smart grid implementation, In: *Smart Grid Security SpringerBriefs in Cybersecurity*, Springer, London, 1–39.

Goldman, A., & Sweet, P. (2008, May 29). Flash! Stealing electricity is risky business. *Las Vegas Sun.* Retrieved February 5, 2011, from ww.lasvegassun.com/news/2008/may/29/flash-stealing-electricity-risky-business/.

Gustafson, M., & Baylor, J. (1989). Approximating the system losses equation [power systems]. *IEEE Transactions on Power Systems, 4*(3), 850–855.

Hagan, M., & Behr, S. (1987). The time series approach to short term load forecasting. *IEEE Transactions on Power Systems, 2*(3), 785–791.

Hale, R.S. (1896). Charging for electric current on the wright demand system—How to adjust rates so that every class of customer shall be profitable to the company. *The Electrical Engineer, 22*(442), 392–393.

Hallberg, J.H. (1905a). Theft of current: How to detect, prosecute and prevent I. *Electrical World and Engineer, 45*(17), 794–796.

Hallberg, J.H. (1905b). Theft of current: How to detect, prosecute and prevent II. *Electrical World and Engineer, 45*(19), 884–886.

Hand, D.J. (1998). Data mining: Statistics and more? *The American Statistician, 52*(2), 112–118.

Haskins, C.D. (1897). Electric metering from the station standpoint. *Transactions, American Institute of Electrical Engineers, 14*(1), 265–274.

Heydt, G. (2010). The next generation of power distribution systems. *IEEE Transactions on Smart Grid, 1*(3), 225–235.

Jain, A.K., Duin, R.P., & Mao, J. (2000). Statistical pattern recognition: A review. *IEEE Transactions on Pattern Analysis and Machine Intelligence, 22*(1), 4–37.

Jiang, R., Lu, R., Wang, Y., Luo, J., Shen, C., & Shen, X. (2014). Energy-theft detection issues for advanced metering infrastructure in smart grid. *Tsinghua Science and Technology, 19*(2), 105–120.

Kelly, E. (1986). An adaptive detection algorithm. *IEEE Transactions on Aerospace and Electronic Systems, 22*(1), 115–127.

Kim, V. (2011, September 17). *Father and daughter burned in alleged electrical theft.* Retrieved September 19, 2011, from LA Times: http://latimesblogs.latimes.com/ lanow/2011/09/father-daughter-burns.html.

Koitzsch, K. (2017). Data visualizers: Seeing and interacting with the analysis. *Pro Hadoop Data Analytics*, Apress, Berkeley, CA, 179–200.

Krishna, V.B., Iyer, R.K., & Sanders, W.H. (2015). Arima-based modeling and validation of consumption readings in power grids. *International Conference on Critical Information Infrastructures Security*, 199–210.

Lee, W., & Stolfo, S. (1998). Data mining approaches for intrusion detection. *USENIX Security Symposium*, San Antonio, TX, 79–93.

Mannila, H. (1996). Data mining: Machine learning, statistics, and databases. *Eight International Conference on Scientific and Statistical Database Management*, 1–8.

McDaniel, P., & McLaughlin, S. (2009). Security and privacy challenges in the smart grid. *IEEE Security & Privacy Magazine*, 7(3), 75–77.

McLaughlin, S., Holbert, B., Fawaz, A., Berthier, R., & Zonouz, S. (2013). A multi-sensor energy theft detection framework for advanced metering infrastructures. *IEEE Journal on Selected Areas in Communications*, 31(7), 1319–1330.

Monedero, I., Biscarri, F., León, C., Guerrero, J.I., Biscarri, J., & Millán, R. (2010). Using regression analysis to identify patterns of non-technical losses on power utilities. In: Setchi R., Jordanov I., Howlett R.J., Jain L.C. (eds.), *Knowledge-Based and Intelligent Information and Engineering Systems. KES 2010. Lecture Notes in Computer Science*, Springer, Berlin, Heidelberg, Vol 6276, 410–419.

Morton, H.D. (1933, October 20). Protective system for electric meters. *US Patent No. 2,019,866*.

Nagi, J., Mohammad, A.M., Yap, K.S., Tiong, S.K., & Ahmed, S.K. (2008). Non-technical loss analysis for detection of electricity theft using support vector machines. *2nd IEEE International Conference on Power and Energy (PECon)*, 907–912.

Nesbit, B. (2000). Thieves lurk, the sizable problem of stolen electricity. *Electrcial World*, 214(5), 31–35.

Nizar, A.H., Dong, Z.Y., & Wang, Y. (2008). Power utility nontechnical loss analysis with extreme learning machine method. *IEEE Transactions on Power Systems*, 23(3), 946–955.

Northeast Group. (2014). *Emerging Markets Smart Grid: Outlook 2015*. Washington, DC: Northeast Group.

Preece, W.H. (1882). Electric lighting at the Paris exhibition. *Van Nostrand's Engineering Magazine*, 26, 151–163.

Seger, K.A., & Icover, D.J. (1988). Power theft the silent crime. *FBI Law Enforcement Bulletin*, 57, 20–25.

Siegel, D. (2017). *Dennis Siegel*. Retrieved June 10, 2017, from http://dennissiegel.de/.

Singhal, S. (1999). The role of metering in revenue protection. *IEE Metering and Tariffs for Energy Supply Conference*, Birmingham, UK.

Smith, T.B. (2004). Electricity theft: A comparative analysis. *Energy Policy*, 32, 2067–2076.

Sreenivasan, G. (2011). *Power Theft*. New Delhi: PHI Learning Private Limited.

Stokes, J.H., Clark, J.I., & Maxwell, C.E. (1986, January 21). Anti-energy diversion system for electric utility meters, *US Patent No. 4,565,995*.

Suriyamongkol, D. (2002). Non-technical losses in electrical power systems. MS Thesis: Ohio University.

Tan, H., Lee, C., & Mok, V. (2007). Automatic Power Meter Reading System Using GSM Network. *International Power Engineering Conference (IPEC)*, 465–469.

Viegas, J., Esteves, P., Melício, R., Mendes, V., & Vieira, S. (2017). Solutions for detection of non-technical losses in the electricity grid: A review. *Renewable and Sustainable Energy Reviews*, 80, 1256–1268.

von der Gracht, H., Salcher, M., & Graf Kerssenbrock, N. (2016). *The Energy Challenge*. München: Redline Verlag, Münchner Verlagsgruppe GmbH. Retrieved from http://webcache.googleusercontent.com/search?q=cache:E0GiNRmmXQwJ; www.uta.edu/faculty/jcchiao/Press_release_8/151113_KPMG/the-energy-challenge.pdf+&cd=1&hl=en&ct=clnk&gl=us.

Wade, H.R. (1955). Kansas city service drop obviates theft, tree problems, contacts. *Electrical World*, 110–111.

Wang, X., Liang, Q., Mu, J., Wang, W., & Zhang, B. (2015). Physical layer security in wireless smart grid. *Security and Communication Networks, 8*(14), 2431–2439.

Ward, M.R., Bihl, T.J., & Bauer, K.W. (2014). Vibrometry-based vehicle identification framework using nonlinear autoregressive neural networks and decision fusion. *IEEE National Aerospace and Electronics Conference (NAECON)*, Dayton, OH, 180–185.

Weslowski, J. (1976). Utilities launch assault to halt theft of power. *Electric Light and Power, 54*(10), 25–26.

Wilson, R.L. (1988, October 18–20). Utility revenue protection. *APPA Accounting, Finance Rates & Information Systems Workshop*.

Wu, X.K., Ghosh, J., Yang, Q., Motoda, H., McLachlan, G., Ng, A., & Zhou, Z. (2008). Top 10 algorithms in data mining. *Knowledge and Information Systems, 14*(1), 1–37.

Yip, S.C., Tan, C.K., Tan, W.N., Gan, M.T., & Bakar, A.H. (2017). Energy theft and defective meters detection in AMI using linear regression. *IEEE International Conference on Environment and Electrical Engineering and 2017 IEEE Industrial and Commercial Power Systems Europe (EEEIC/I&CPS Europe)*, 1–6.

Zhang, G.P. (2007). Avoiding pitfalls in neural network research. *IEEE Transactions on Systems, Man, and Cybernetics, Part C (Applications and Reviews), 37*(1), 3–16.

7

Unit Commitment Control of Smart Grids

Salam Hajjar

Marshall University

CONTENTS

7.1 Introduction 125
7.2 Renewable Energy Resources 126
    7.2.1 Wind Power 126
        7.2.1.1 Wind Power Generation 127
        7.2.1.2 Wind Turbine Control 127
    7.2.2 Solar Power 128
        7.2.2.1 Solar Panels 129
        7.2.2.2 Solar Panel Capacity 129
        7.2.2.3 Solar Panel Efficiency 130
        7.2.2.4 Solar Panel Power Generation Density 130
        7.2.2.5 Power Collected vs. Energy Collected 130
7.3 The Unit Commitment Problem 131
    7.3.1 Illustrative Example 131
    7.3.2 The Unit Commitment Problem 133
7.4 A Multi-agent Architecture 133
    7.4.1 Smart Grid Using Multi-Agent Model 134
    7.4.2 Agent's Profile 135
    7.4.3 Decision-Making Method 136
    7.4.4 Storing and Selling Extra Power Procedure 138
7.5 Illustrative Example 139
7.6 Conclusions 140
References 140

7.1 Introduction

Renewable energy resources, such as wind and solar, have been considered as non-negligible sources of backup energy in the recent decades. While these resources are free and ceaseless sources of energy that can be used to generate electrical power for human demands, they are unpredictable in nature. The unit commitment (UC) problem is defined in the literature of electrical power production as the problem of producing power by collaboration of renewable generators in order to achieve consumer demand. In earlier decades, primarily conventional energy sources, e.g., fossil fuel, were used to generate electrical power due to the predictability of power they can provide. In contrast, renewable energy sources can be unpredictable due to their reliance on variable natural sources, e.g., sunlight which can vary day-to-day due to cloud cover.

To better understand the uncertainty issue, one may think about the weather conditions. For example, a cloudy or rainy day may cause limited generation of renewable power, whereas a sunny and windy day may produce an overabundance of power. However, a sunny day could potentially reach almost 90% of the total power consumed in a specific geographic area. In both cases, conventional generation sources have likely produced the same amount of power. Of course, this is a rare case and cannot be taken as a reference. However, the change in the power generation certainly raised a flag directing to the importance to keep the energy production rate in scale with the energy consumption. In this chapter, we introduce the renewable energy resources and explain a centralized algorithm that solves the UC problem for smart power grid containing renewable power generating components (solar panels and wind turbines).

7.2 Renewable Energy Resources

In this section, we provide a brief idea about the renewable power energy from a mechanical point of view. In total, the energy from the sun reaching or planet in one hour is greater than what is used by everyone in the world in one year. This shocking information turned the researchers toward the renewable energy (Munroe and Shepherd, 1981). Contemporarily, no one can neglect the important role that the renewable energy resources play in our daily life. Many devices are now using solar power to recharge batteries, heat water, run electric devices, etc. However, we cannot depend completely on the renewable resources to generate power for two main reasons: the first one is the production rate of renewable resources and is small compared to the conventional resources. The second reason is the uncertainty of availability of these resources (Neij, 1999; Nilsson and Bertling, 2007). Thus, in a UC problem, where a utility is required to satisfy the demands of its clients, one cannot exclude the conventional power resources, and renewable resources can be used as backup and supporting units only.

7.2.1 Wind Power

Wind is a natural kinetic energy source which is largely produced by the sun and other natural conditions, such as the differential heating of the atmosphere. Additional factors to wind availability and strength include (1) the rotating movement of the Earth and (2) local geographical features, such as mountains Wind's speed can reach 30 mph and may be dangerous when it exceeds 60 mph and can cause damages to human and belongings. However, this wind, when collected and manipulated by modern wind turbines, can be used to generate electricity.

7.2.1.1 Wind Power Generation

Electrical power from wind is widely generated using wind turbines (Shepherd and Shepherd, 2003; Shepherd and Zhang, 2011). A wind turbine is a device which converts the kinetic energy of the wind into rotational energy which drives an electric generator (Shepherd and Shepherd, 2003; Shepherd and Zhang, 2011). Typical wind turbines consist of two to three blades attached to low speed rotating shaft though a hub. The rotating shaft spins in the same speed as the blades, which is usually 7–10 rpm. To produce electric power from the mechanical spinning, it is necessary to increase the rotating speed to few hundreds of turns per minute; thus, a gearbox is mounted around the shaft to transfer the speed to a high-speed shaft. The gearbox is also used to control the speed and protect the turbine against high speeds. The high-speed shaft is connected to a mechanical power generator that converts the snipping movement into a direct current (DC; Shimizu et al., 1996; Shepherd and Zhang, 2004). The amount of power a wind turbine can produce depends on the size of the turbines and large turbines are generally able to produce about 2 MW of power. Wind farms, which are a cluster of possibly a few hundred wind turbines, may reach around 180 MW of power, depending on its size and the number of active turbines.

7.2.1.2 Wind Turbine Control

The main goal of the wind turbine is to generate the maximum amount of power without damaging the device. The electric power generated by a wind turbine can be calculated by Ohm's law as follows:

$$P = \frac{V^2}{R} \tag{7.1}$$

where V is the voltage created and R is the rotor part's resistance. It is obvious that decreasing the resistance increases the power generated by the turbine. However, changing the blades' pitch also plays an important role in controlling the turbine. Finding the best pitch for the blades depends on the length, the shape, and the material of the blade. It can be indicated that two pitches will prevent the turbine of working, which are when the pitch is equal to 0° and 90°. In the first case, the wind will be orthogonal to the blades and in the second one the wind will be parallel to the blades which will make the effect of the wind over the blades equal to zero (Kreyszig, 2011).

Various factors exist to impact the performance of a wind turbine. There exist some automatic control systems to adjust the blades pitch and move the blades' angle to the best position regarding the wind direction and speed. Other factors affect the performance of the wind turbine, such as (1) the height of the turbine tower, the higher the turbine is the more it can capture wind, and (2) the material and number of blades. The more blades a turbine has, the better it can perform. However, the cost of a blade of a commercial wind turbine can reach thousands of dollars, and depends on its length, fabrication material, weight, sustainability, and other factors. The recommended number of blades is 3, as a turbine with three blades can produce an amount of power close to what a four-blade turbine can provide. However, for a two-blade turbine in order to provide a close amount of power as a three-blade one it needs to turn much faster than the three-blade one, which makes of the turbine very noisy and requires extra costs to ensure its safety.

7.2.2 Solar Power

The solar power generation is based on the photovoltaic (PV) technology, where solar cells or PV cells, as shown in Figure 7.1a, convert sunlight directly into electricity. The electrical circuit diagram of a PV cell is illustrated in Figure 7.1b. These cells are made up of silicon, which is a semiconductor element that is able to convert the sunray elements, known as photons, to a DC. The silicon cell when exposed to light creates a DC (Shepherd and Shepherd, 2003). A typical silicon PV cell illustrated in Figure 7.1 is composed of two layers: (1) a phosphorus-doped (N-type) silicon and (2) a layer of boron-doped (P-type) silicon. Because of the difference of charges embedded in each layer, an electrical field is generated around the phosphorous layer and makes what is called the P–N junction (Shepherd and Shepherd, 2003). When light hits the surface of the solar cell, this electrical field enforces the light-stimulated electrons to move together in certain direction and certain speed, which results a flow of DC once the solar cell is connected to an electrical load (Shepherd and Shepherd, 2003).

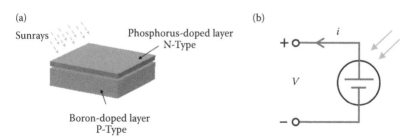

FIGURE 7.1
Conceptualizations of solar cell: (a) P–N junction conceptualization and (b) circuit diagram.

7.2.2.1 Solar Panels

An assembly of solar cells electrically connected forms a solar panel. Each solar cell is rated by the DC it can produce if exposed to standard test conditions (STC). Typical solar cell's rate ranges from 90 to 350 W. Because a single solar cell is limited in amount of power generated, most installations contain tens of cells in a solar panel. An example of a solar panel is presented in Figure 7.2. Here, the solar panel contains many wired solar cells along with electronics and a housing to support the cells. DC travels through the wires to carry the electricity produced in the cells to a junction unit where the panel is attached to a grid. It is obvious that the more cells involved in a panel, the more power can be generated. Thus, the size of a solar panel does matter in energy production. One typical size for industrial solar panels is 39″×65″, which contain 60 solar cells on average.

7.2.2.2 Solar Panel Capacity

The power produced by the solar panel call the panel capacity can be calculated by the following equation:

$$P_t = n \times P_{cell} \tag{7.2}$$

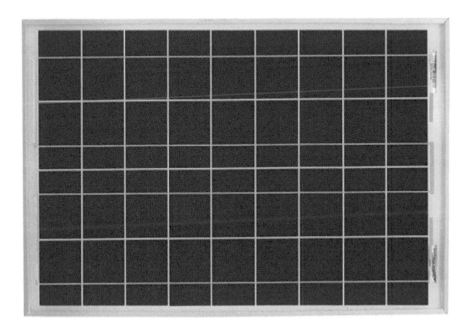

FIGURE 7.2
Example of a representative solar panel.

where P_t is the total power produced by the solar panel, P_{cell} is the power generated by the cell, and n is the number of cells mounted on a solar panel. The panel capacity calculated above represents the maximum amount of power that an ideal solar panel can generate. But, in reality, the solar power production is also affected by the efficiency of the panel.

7.2.2.3 Solar Panel Efficiency

Developments in solar cell technology heavily focus on the chemistry and physics surrounding the process and the materials used (Shepherd and Shepherd, 2003; Green et al., 2015). Late in the 1950s, the industrial solar panels of a size 40″×60″ had an efficiency of 6% and were able to generate about 20 W, a power sufficient to turn on only one electric 20 W bulb. However, gains in efficiency [currently about 20%–30% per (Green et al., 2015)] have resulted in panels of the same size being able to produce about 265 W, which can turn on 12 electric 20 W bulbs. Further recent advances include laboratory developments of 40% efficiency cells (Green et al., 2015); however, these are considerably more expensive than the typical ones available but show further promise for future solar utility. The efficiency of any system is defined as the ration of the system's output to its input (Melhem, 2013). Thus, the efficiency of a solar panel is calculated as follows:

$$Eff = \frac{P_{out}}{P_{in}} \tag{7.3}$$

where P_{in} is the amount of power the panel receives from the sun and P_{out} is the amount of power the panel provides to the user.

7.2.2.4 Solar Panel Power Generation Density

The maximum power generation density for a solar panel is an important property of a panel, denoted by P_d, measured by Watts by meter squared, and is calculated as follows:

$$P_d = \frac{P_t}{A} \; (W/m^2) \tag{7.4}$$

where P_t is the total panel capacity and A is the panel area given by (length × width) (Shepherd and Shepherd, 2003). To facilitate sizing, online calculators are available (Energy Groove, 2017).

7.2.2.5 Power Collected vs. Energy Collected

Solar panels are also rated by the power they can collect from sun, which is a value measure by W/m^2. The sun radiation is usually measured by W/m^2.

The power collected by the solar panel is equal to the amount of solar power radiation times the panel efficiency. It is calculated as follows:

$$P_{collected} = S \times Eff \quad (W/m^2) \tag{7.5}$$

where S is the power provided by the sunray.

One must keep in mind that the power collected from the sun cannot exceed in practice the power generation density P_d calculated in Equation (7.4).

The energy collected by a solar panel is equal to the power collected by the panel during certain amount of time. It is denoted by $E_{collected}$ and is calculated as follows:

$$E_{collected} = P_{collected} \times T \quad (W \cdot h/m^2) \tag{7.6}$$

A panel of a capacity of 200 W rating will provide 200 watt-hours of electricity. One can get about 1 kWh by using five of such panels. If these panels work for five sunny hours per day, they produce 5 kWh. Per month, they can produce $5 \times 30 = 150$ kWh. A typical residential apartment of 1,000 square feet in the USA consumes around 600 kWh per month. Thus, 20 solar panels of 200 W rate would be required to power one single average apartment using only solar energy.

Since solar power is available during the day, but demand for its use might be at night, it is of a high importance to mention that the power generated by the solar panels is generally saved in deep cycle batteries. This specific type of batteries is used to because it enjoys the ability to make deep prolonged and repeated discharge of the battery which is typical for the solar energy. Since the load devices usually need alternating current (AC) to work, a DC/AC inverter is used to convert the DC power generated into an AC one (Shepherd and Shepherd, 2003).

7.3 The Unit Commitment Problem

Due to the unreliable nature of solar power, methods must be used to optimize its use. The UC problem facilitates using renewable energy power in conjunction with other power sources by monitoring and tracking production and demand.

7.3.1 Illustrative Example

For example, we will consider a solar panel of a size 24″×21″ with a 50-W capacity and efficiency of 40%. One further consideration is that the user of this panel requires an average of 10,000 W per day.

Table 7.1 presents solar radiation values for this panel along with times (2 h periods considered) and the resultant power and energy collected. The panel is assumed to be exposed to the sun during a day, with variation due to time of day. Calculating using the equations mentioned in Section 7.4, one can compute the following values: (1) the power generation density, (2) the total power collected, (3) the total energy collected by the panel if the panel is displayed to the sun that is providing various radiation rate as shown in the table, (4) what is the panel area (m²) required to serve a load demand of 10,000 W · h required per day, and (5) How many panels are required to cover the client demand.

1. For calculation purposes, one computes values in the following order: Calculate the power density we first need to calculate the panel's area.
2. $A = 24 \times 21 = 504''^2$. We convert the area into m² because the power and energy collected are measured in m², then, $A = 0.325\,m^2$.
3. Thus, $P_d = 50/0.325 = 153.8\,W/m^2$.
4. The power collected is calculated by Equation (7.4).
5. The energy collected is calculated by Equation (7.5).

The energy collected by the panel during the day is the accumulation of the energy collected from 00:00 am to 23:59 am, for this example the total energy collected was 3,399.2 W · h/m² and the total power collected = 1,699.6 W/m².

Now we can determine the appropriate area of panels we would need to satisfy the customer's 10,000 W demand. Here we can divide the load of

TABLE 7.1

Solar radiation and corresponding power and energy collected

From	To	Increment Duration	Solar Radiation (W/m²)	Power Collected (W/m²)	Energy Collected (W·h/m²)
0:00	2:00	2	0	0	0
2:00	4:00	2	0	0	0
4:00	6:00	2	0	0	0
6:00	8:00	2	150	60	120
8:00	10:00	2	550	220	440
10:00	12:00	2	850	340	680
12:00	14:00	2	1,020	408	816
14:00	16:00	2	904	361.6	723.2
16:00	18:00	2	600	240	480
18:00	20:00	2	175	70	140
20:00	22:00	2	0	0	0
22:00	23:59	2	0	0	0

10,000 W by the total power collected per day by this panel. To determine the number of panels needed, the result is 5.88 m^2 of area would then be divided by the area of each panel, 5.88/0.325 = 18.1. Thus, 19 total panels would be needed since fractional panels are not feasible.

7.3.2 The Unit Commitment Problem

In the former sections, we provided the reader with a general idea about the renewable energy sources and generation. In the coming sections, we discuss the problem of UC from the control point of view. Before diving in the control process it is of important to indicate that pure costs of power generation are supposed to be standard and fixed to all power generation units; however, the production, installation, and transportation techniques vary from one utility to another resulting the change of power prices charged on the client (Conejo, Plazas, Espinola and Molina, 2005; Livel, 2010). It is of important to highlight that client demand also varies regarding the geographical location of the served site, the time of the day and the year. The information required to forecast the production and the consumption are collected in a time frame of one day ahead. To compute the consumption forecast a day ahead, the data should be collected hourly or each 30 min of the client demand, the weather state (temperature, wind speed, sky clarity, etc.). As discussed in Chapters 2 and 3, some sophisticated forecasting methods using artificial intelligence techniques can be employed in order to compute the production forecast for the next day (Tan, 2002; Dubost et al., 2005; Sahay and Tripathi, 2014).

7.4 A Multi-agent Architecture

The multi-agent system is a platform used to simplify the communication among a group of collaborating members, named agents, to achieve a certain mission or realize satisfy a certain need (der Hoek and Wooldridge, 2008). Java Development (JaDe) (Bellifemine et al., 2001) is one framework that can be used to implement multi-agent systems. Social network websites, such as Facebook and Twitter, and commercial websites, such as EBay and Amazon, can be seen as multi-agent systems, as they are gathering many communicating participants who share common privileges and getting connected through a network for a specific goal. In such a framework, agents can be active or inactive; they can also be of different categories and have different rights and privileges within the network. For example, categories could be admin, buyer, seller, coordinator, etc. Some multi-agent systems are built out of smaller subsystems that communicate through a representative agent of each subsystem. Figure 7.3 shows an example of a multi-agent system, and Figure 7.4 shows an example of two communicating subsystems forming a

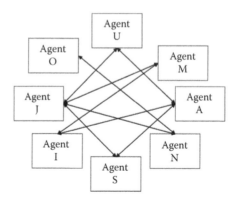

FIGURE 7.3
Multi-agent architecture.

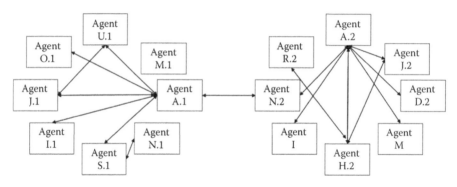

FIGURE 7.4
Multi-agent system built of two communicating subsystems.

global multi-agent system. Employing such architecture in a smart grid modeling is very useful, as it simplifies the communications among agents and the decision-making process for the agents with some controlling responsibilities in the system.

7.4.1 Smart Grid Using Multi-Agent Model

A smart grid can be modeled as a multi-agent system (Pipattanasomporn et al., 2009), where agents are divided into three main categories: (1) producers, (2) consumers, (3) storage elements (batteries), (4) external market, and (5) controllers. In real-life application, an agent can belong to more than one category, but cannot participate in a process under more than one category in the same time. For example, we can think of an electrical vehicle EV as an agent that can be a consumer of power when it is charging at a station for some time; however, that same EV can act as a producer when it is discharging at a station for some other times (e.g., Mwasilu et al., 2014). For simplicity, and consistent with (Hajjar et al., 2015), we consider in this chapter that an agent belongs to only one category for all times.

7.4.2 Agent's Profile

Each agent in the grid has a private profile, contains all the details about this agent, such as Identification number, passcodes, location, production or consumption rates, availability status, computation method, price, power needed, power available, etc., and a public profile to share with other agents in the grid, that contains only the information that the agent wants to share on the grid.

For convenience, we will consider that a smart grid contains a set of (1) conventional power producers, such as gas turbines, nuclear plant or a hydraulic turbine, (2) renewable power producers, such as wind turbine or PV panel, (3) consumers, which can be building, factory, house, etc., and (4) a unique controller that manages the selling and buying process in the grid (Hajjar et al., 2015). The agents have the following public profiles:

- Conventional and renewable power producer profiles:
  - Agent's ID
  - Agent type: conventional or renewable
  - Availability status: available/unavailable
  - Selling price: amount measured in U.S. dollars
  - Energy available for selling: amount measured in Watts per hour
- Consumer profile called request:
  - Agent's ID
  - Power demand: a value measured in watts.
  - Buying price: the maximum price a client can pay, a value measured in U.S. dollars.
- Storage element public profile:
  - Agent's ID
  - State of charge (SoC)
  - Full capacity
  - Storage price

A power producer profile is presented as follows:

$$Producer\_profile = (agent\_ID, status, selling\ price, power\ for\ selling) \quad (7.7)$$

whereas a consumer profile is presented as follows:

$$Consumer\_profile = (agent\_ID, demand, buying\ price), \qquad (7.8)$$

The storage agent profile is presented as follows:

$$Storage\_profile = (agent\_ID, SoC, full\ capacity, storage\ price)\,\text{m}^2 \qquad (7.9)$$

For an example of the use of this notation, a gas turbine agent offer at any time of the day is given as follows: (GT #11, available, $4, 1000 W·h), a PV agent offer at 3 pm in a sunny day is given as follows: (PV #123, available, $0.5, 200 W·h), and a rest area stop on a highway has a request profile at 3 pm in a vacation day is given as follows: (consumer #123, 1,000 W, $2.00).

One assumption employed herein is that the market has a non-limited capacity and can always accept the sold power. When considering the controller profile, the controller does not have a public profile to share with other agents in the grid. It has two types of UC plans. The first is the UC Producer Plan (UCPP), and second is the UC Consumer Plan (UCCP). A specific UCPP will be sent to each producer contains the amount energy that will be bought from the agent named *power_to_sell* and its next working status (ON or OFF). Similarly, a specific UCCP will be generated and sent to each consumer including the set of all the producers that will cover the demand and the total purchase cost.

A simplifying condition in this grid states that consumers and producers can communicate only with the controller, they cannot communicate between each other; however, consumers can communicate among each other and producers (conventional and/or renewable) can communicate among each other too. The controller is supposed to receive offers and requests from all agents and provide a UC plan containing the best offer for each client and the working plan for the producers. In other words, the controller makes the decision of turning ON and OFF the producing agents whenever needed. Figure 7.5 illustrates a high-level representation of a multi-agent smart grid.

7.4.3 Decision-Making Method

The decision-making method explained in this section is high-level abstraction of the method presented in Hajjar et al. (2015). The controller generates a UC plan based on the values received and the algorithm illustrated in Figure 7.5 with steps enumerated in Figure 7.6. In step 1 of Figure 7.5, the controller collects the offers from the producing agents. Each offer includes the agent's profile that contains the producer's ID, its availability, the amount of energy it can provide in the coming period, and the selling price. The controller also receives the demand requests from the

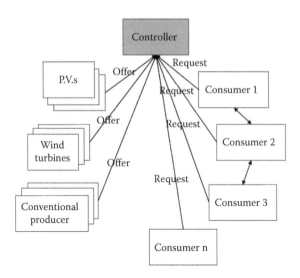

FIGURE 7.5
Simplified multi-agent model of a smart grid.

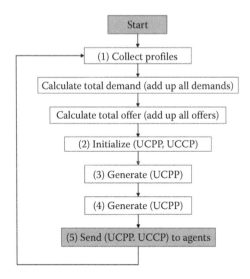

FIGURE 7.6
UC plan generating algorithm.

consumers, containing the energy needed and the buying price. In step 2 of Figure 7.5, the controller initializes the UCPP of each producer, by setting the working status to OFF and the amount of energy bought equals zero. The controller also initializes the UCCP of each consumer, by setting the list of serving agents to empty and the purchase cost to zero. In steps 3 and 4 of Figure 7.5, the controller generates for each producer a UCPP and a UCCP for each consumer using the strategies shown below. In step 5, the plans are sent from the controller to the consumers and producers. If, after covering the demand of consumer, there remains extra produced energy then the controller decides whether to stoke the energy in batteries for the next day or sell it to the external market regarding the SoC of the batteries and the market price.

Within the "Generate (UCPP)" action seen in Figure 7.6, the following operations are performed:

1. Read offer
2. If (*status* is available) & (*selling price* < *buying price*) & (*power to sell* > 0) then
   a.   set *work_status* to ON;
   b.   if (*demand* < *power_to_sell*) then *power_to_sell* = *power_to_ sell* – *demand*
        else *power_to_sell* = *power_to_sell* (the agent is selling its total production of energy);
   c.   update demand: *demand* = *demand* – *power_to_sell*;

Similarly, within the "Generate (UCCP)" action, the following operations are performed:

1. Read *demand*
2. For each producer (test producers in order) If (*buying_price* > *selling_price*) then
   a.   add producer to the list of serving producers;
   b.   set *cost* = *cost* + *selling_price*;
   c.   update demand; *demand* = *demand* – *power_to_sell*;
3. move to next producer.

The UCCP procedure at step 2 within Figure 7.5 will be executed over all the producing agents in a specific order, to appeal the cheaper producers in first place: (1) PVs, (2) wind turbines, (3) nuclear, (4) gas turbine, and (5) hydraulic turbine.

7.4.4 Storing and Selling Extra Power Procedure

At each sampling period, the controller tests if there exists extra energy not consumed by the local consumer. If there is extra energy, then the controller commands the producers to store the energy in the available batteries for a certain storage cost calculated in Equation (7.7) depending on the batteries' capacity and SoC:

$$Storage\_cost = storage\_price \times extra\_energy \quad (\$) \qquad (7.10)$$

Once the batteries are fully charged, the remaining energy will be sold to the external market.

$$Sold\_energy = extra\_energy - stored\_energy \quad (W \cdot h) \qquad (7.11)$$

The gain of purchase is calculated as follows:

$$Gain = (market\_price \times extra\_power) - storage\_cost \quad (\$) \qquad (7.12)$$

7.5 Illustrative Example

Figure 7.7 illustrates the energy production and consumption forecast data that are collected on November 20, 2017 for the entire area of France.[*] As seen in Figure 7.7, collected data included samples at 15 min intervals from 0:00 until 23:45 pm for gas, nuclear, wind, solar, and hydraulic sources, and total consumption.

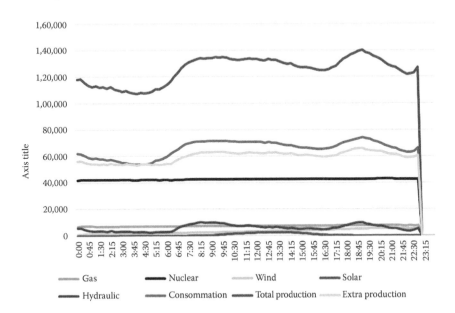

FIGURE 7.7
Energy production/consumption forecast for France on November 20, 2017.

---

[*] Data collected from the website of French Electricity Transportation Grid. www.rte-france.com/en/eco2mix/eco2mix-telechargement-en.

As an example of the UC problem, we can take the profiles at 3 pm and compute the difference between the total consumption and production as 60786 MW·h. The controller agent can be improved to make the decision of selling the produced energy to an external market and calculate the grid's gain. For example, if we assume the given the SoC of batteries in the system equal to 100%, and the price of market selling is about 12 cents/kWh, then the total gain is equal to $7,2943,200.00.

7.6 Conclusions

In conclusion, one can recognize that the multi-agent framework could cut down the complexity of the control process of a smart grid. Organizing the control process by proposing clear steps such as specifying the agents' categories, defining the profile of each sort of agents, and controlling the messages' exchange process among all agents resulted an organized collaboration and prevented a chaotic behavior inside the grid. We can affirm that the multi-agent framework can be used for small grids as well as for very large ones. The only detail the designer must worry about is defining the agents and carefully filling the profile of each agent.

References

Bellifemine, F., A. Poggi, and G. Rimassa. 2001. Developing multi-agent systems with JADE. In *International Workshop on Agent Theories, Architectures, and Languages.* Springer, Berlin, Heidelberg, 89–103.

Conejo, A.J., M.A. Plazas, R. Espinola, and A.B. Molina. 2005. Day-ahead electricity price forecasting using the wavelet transform and ARIMA models. *IEEE Transactions on Power Systems, 20*(2), 1035–1042.

Van der Hoek, W., and M. Wooldridge. 2008. *Multi-Agent Systems, Foundations of Artificial Intelligence,* Elsevier, 3, 887–928.

Dubost, L., R. Gonzalez, and C. Lemaréchal. 2005. A primal-proximal heuristic applied to the french unit-commitment problem. *Mathematical programming, 104*(1), 129–151.

Energy Groove. n.d. Solar Panel Calculator. Accessed December 20, 2017 www.energygroove.net/energy-cost/solar-panel-calculator/.

Green, M.A., K. Emery, Y. Hishikawa, W. Warta, and E.D. Dunlop. 2012. Solar cell efficiency tables (Version 39). *Progress in Photovoltaics: Research and Applications,* 20, 12–20.

Hajjar, S., A.I. Bratcu, and A. Hably. 2015. A day-ahead centralized unit commitment algorithm for a multi-agent smart grid. *Federated Conference on Computer Science and Information Systems,* Lodz, Poland, 265–271.

Kreyszig, E. 2011. *Advanced Engineering Mathematics*. John Wiley & Sons, Hoboken, NJ.

Lively, M.B. 2010. Short Run Marginal Cost Pricing for Fast Responses on the Smart Grid. In *Innovative Smart Grid Technologies (ISGT)*, IEEE, New York, 1–6.

Melhem, Z. 2013. *Electricity Transmission, Distribution and Storage Systems*. Elsevier, New York, NY.

Munroe, M., and W. Shepherd. 1981. An assessment of solar energy availability in different regions of the solar spectrum. *Solar Energy*, 26(1) 41–47.

Mwasilu, F., J.J. Justo, E.K. Kim, T.D. Do, and J.W. Jung. 2014. Electric vehicles and smart grid interaction: A review on vehicle to grid and renewable energy sources integration. *Renewable and Sustainable Energy Reviews*, 34, 501–516.

Neij, L. 1999. Cost dynamics of wind power. *Energy*, 24(5), 375–389.

Nilsson, J., and L. Bertling. 2007. Maintenance management of wind power systems using condition monitoring systems—Life cycle cost analysis for two case studies. *IEEE Transactions on Energy Conversion*, 22(1), 223–229.

Pipattanasomporn, M., H. Feroze, and S. Rahman. 2009. Multi-agent systems in a distributed smart grid: Design and implementation. Power Systems Conference and Exposition. *PSCE'09*. IEEE/PES, 1–8.

Sahay, K.B., and M.M. Tripathi. 2014. Day ahead hourly load forecast of PJM electricity market and ISO New England market by using artificial neural network. *Innovative Smart Grid Technologies Conference (ISGT)*. IEEE PES.

Shepherd, W., and D.W. Shepherd. 2003. *Energy Studies*. World Scientific, Hackensack, NJ.

Shepherd, W., and L. Zhang. 2004. *Power Converter Circuits*. CRC Press, Boca Raton, FL.

Shepherd, W., and L. Zhang. 2011. *Electricity Generation Using Wind Power*. World Scientific, Hackensack, NJ.

Shimizu, Y., M. Takada, H. Fujiwara, Y. Yasui, T. Maeda, Y. Kamada, T. Hori, M. Ishida, and N. Yamada. 1996. Studies on horizontal axis wind turbine electric generation system with variable speed control. *4th International Workshop on Advanced Motion Control*, Mie, Japan.

Tan, B. 2002. Production control of a pull system with production and demand uncertainty. *IEEE Transactions on Automatic Control*, 47(5), 779–783.

8

A New Transformer Differential
Protection Algorithm Based on
Data Pattern Recognition

Ernesto Vázquez Martínez, Héctor Esponda Hernández,
and Manuel A. Andrade Soto
Universidad Autónoma de Nuevo León

CONTENTS

8.1 Big Data and Power System Protection.. 143
8.2 Methods for Differential Protection Blocking 146
    8.2.1 Harmonic Restraint and Harmonic Blocking............................ 146
    8.2.2 Methods Based on Waveform Recognition.............................. 147
8.3 Principal Component Analysis.. 148
8.4 Curvilinear Component Analysis (CCA) 149
8.5 PCA Applied to Discriminate Between Inrush and Fault
    Currents in Transformers 151
8.6 Application of the CCA as a Base for a Differential Protection 154
8.7 System Under Study 157
8.8 Results 160
    8.8.1 Discussion of the Principal Component Analysis Results 160
    8.8.2 Discussion of the Curvilinear Component Analysis Results ...164
8.9 Conclusion 167
References.. 167

8.1 Big Data and Power System Protection

Recent developments in power system measurement and sensor technology, data formats, data management, and storage have given rise to challenges related to the amount of data rapidly increasing in many application domains that should be collected, stored, and analyzed (Kezunovic et al., 2013). Thus, large data sets are becoming common in power systems analysis, protection, and control. The relations between different big data sets and power system functions are shown in Figure 8.1. For example, market and weather data are

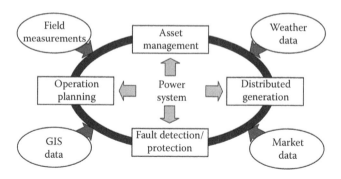

FIGURE 8.1
Relationship between big data sets (field measurements, weather data, Geographic Information Systems data, market data, etc.) and power system functions (asset management, distributed generation, fault detection/protection systems, operation planning).

required to set the generation level of distributed generators and their operation modes. The growth of available information and its impact has caused the fact that this extremely large data set, "big data," has been proclaimed as one of the major challenges of this century (Heskes, 1999). Although the analysis of these large volumes of data often carries the promise to gain new insights into relevant processes and phenomena, it also raises many issues in terms of computing complexity (time and memory consumption), workload distribution, and efficient processing of the results (Bunte and Lee, 2015; Zhang and Yang, 2016).

Protective relaying schemes are designed to minimize the fault clearing time while maintaining the reliability (measured in terms of dependability and security) of the scheme. The challenge is to process sampled values of voltage and current signals (e.g., 128 samples per 60-Hz cycle, in three-phase systems, i.e., 23,040 samples per second) to identify a fault inception in the system.

When a power transformer has an internal fault, it should be disconnected as soon as possible to avoid extensive damage in the equipment, to preserve the stability of the power network and to improve power quality (Phadke and Thorp, 2009). The main protection used in power transformer is the differential protection, which is based on Kirchhoff's laws of electric circuits. This protection scheme has excellent performance to detect internal faults, but it can operate incorrectly due to magnetic saturation in current transformers caused by dc component present in the fault current. The percentage differential principle improves the performance of the differential protection because it makes a comparison between operation and restraint currents, which are calculated as vector addition and subtraction. It allows reducing the negative effect of the saturation in current transformers (Cordray, 1931).

There are some methods to avoid an incorrect operation of differential relays due to inrush currents during transients affecting the magnetic flux of the power transformer. A first solution was to include an intentional time delay in the differential relay, to halt the relay operation for a time interval (16 ms typically). A disadvantage of this method is a delayed tripping for a short circuit during the inrush condition. Another method, widely used, is a harmonic-current restraint or blocking (Malik, et al., 1976; Abniki, et al., 2010). The differential relay self-desensitizes during inrush currents, but the relay can detect short-circuit currents. To do this, the relay distinguishes between inrush current and short-circuit current by the difference in the current's wave shape. The methods based on harmonic-currents do not work reliably in cases where the differential currents show low harmonic content (Guzman, et al., 2002), such as new magnetic steel is used in the core of transformers. Transformer over-excitation is another possible cause of incorrect operation in the differential relay.

Researchers have proposed to use an additional fifth-harmonic restraint (Rahman, 2006), and waveform recognition algorithms to distinguish faults from inrush to improve relay performance (Youssef, 2003; Tripathy, 2012; Ashrafian, et al., 2017).

This chapter describes two new algorithms for power transformer differential protection. They are dimensionality reduction methods based on the principal component analysis (PCA) and curvilinear component analysis (CCA). PCA constitutes a classical linear technique which is widely used due to its simplicity and efficiency. CCA offers a multidimensional scaling which is based on the aim of distance preservation. However, both techniques have the drawback that their capability to capture nonlinearities can be rather limited (Lèe and Verleysen, 2007). The goal is to reduce the dimensionality of the high-dimensional data space formed by the differential currents to find projections resembling the characteristics of the original space as much as possible and ultimately distinguish between internal faults (short circuits inside transformer differential protection zone) from inrush current (transformer connection and over-excitation). Modal analysis and neural networks are used as pre-processing step in PCA and CCA, respectively; a pattern recognition is carried on using different features from the differential current. While for PCA, the three-phase differential current signal is sampled at 8 kHz (128 samples per cycle) and a sliding window of half a cycle is used (64 samples), CCA uses a whole cycle window sampled at 4 kHz (64 samples per cycle). This process is depicted in Figure 8.2. The PCA and CCA algorithms consider current transformers' saturation phenomena and deviations in the power system parameters. Both algorithms were proven by means of time domain simulation of distinct scenarios as variations in transformer load, source impedance, turn ratio, and $B - H$ curve of current transformers. The algorithms distinguish internal short circuits, inrush, and over-excitation conditions in all cases.

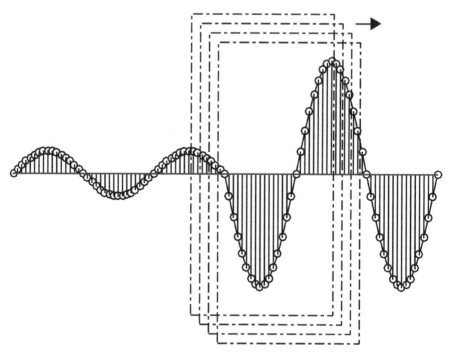

FIGURE 8.2
Sliding window for signal processing.

8.2 Methods for Differential Protection Blocking

8.2.1 Harmonic Restraint and Harmonic Blocking

Harmonic-content-based methods are used to prevent maloperation of differential protection, using the harmonic content of the relay differential current signal. The aim is to increase the restraining current value (harmonic restraint) or to inhibit relay operation (harmonic blocking). It has been shown that the second harmonic component prevails, throughout the harmonic spectrum, during an energizing condition. However, the fifth harmonic component prevails during an over-excitation condition (Guzman, et al., 2002).

In the case of harmonic blocking, the logic states that if the magnitude of the second or fifth harmonic components contained in the differential current exceeds a preset percentage of the fundamental component, then it is a transformer energization or over-excitation condition, respectively, thus blocking the operation of the differential protection to avoid unnecessary disconnection.

The harmonic blocking method will be implemented when any of the following conditions are satisfied:

$$I_{op} < K_2 I_{h2},$$
$$I_{op} < K_5 I_{h5}, \tag{8.1}$$

where K_2 and K_5 are the values to be used as predefined reference values, in percent, to start or block the relay, and I_{h2} and I_{h5} are the magnitude of the second and fifth harmonic component, respectively.

For the case of harmonic restraint, it is required the following condition to be fulfilled:

$$I_{op} > I_R \times SLP + K_2 I_{h2} + K_5 I_{h5}, \tag{8.2}$$

where I_R is the restraining current and SLP is the relay slope characteristic. The effect obtained is to increase, by a certain percentage, the original relay characteristic, thus reducing the operating region and increasing the restrain region.

8.2.2 Methods Based on Waveform Recognition

Other techniques used to prevent relay maloperations due to inrush currents and over-excitation conditions are those based on the direct recognition of the distortion of the differential current waveform. Figure 8.3 shows two waveforms, corresponding to a condition of energization and a short circuit, respectively.

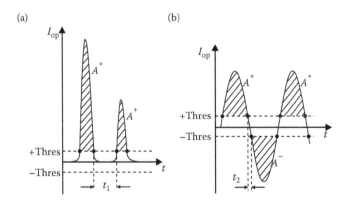

FIGURE 8.3
Inhibition method by waveform recognition for (a) an energizing condition and (b) a short-circuit fault.

The method consists of detecting differential current intervals close to zero, comparing it with two thresholds (abbreviated Thres): a positive one (+Thres) and a negative one (–Thres). The output pulses generated by those thresholds have a duration of t_1 seconds for the energization case, and t_1 seconds for the short-circuit case. These thresholds are compared to a certain time reference, allowing the discrimination between short circuits and energization conditions.

Despite the good performance they have shown, these desensitizing methods have a major drawback: in situations where the waveform of the inrush current has a high degree of symmetry (a well-defined sinusoidal) or when the fault current is highly asymmetric, these methods tend to operate incorrectly showing difficulties to correctly distinguish whether it is a short circuit or energization condition.

8.3 Principal Component Analysis

PCA is a powerful multivariate statistical technique employed to transform a set of variables along orthogonal axes or to reduce the dimension (number of variables) (Dillon and Goldstein, 1984; Jolliffe, 2002). Proposed by Pearson (1901), PCA was further developed by Hotelling (1933) and aims to reduce a database with many variables to a lower dimensional space while losing as little information as possible (Kiliç et al., 2009). The main utility of PCA is that it allows the study of multidimensional phenomena where some or many of the included variables are correlated with each other in varying degrees.

The PCA consists in applying a transformation which results in an orthogonal rotation of the original data space. This new set of orthogonal vectors is determined so they contain the most information of the data variance. This method aims to find linear combinations of variables ($Y = [y_1, y_2,..., y_p]$) representing certain multidimensional phenomenon ($X = [x_1, x_2,..., x_p]$), with the property that they exhibit maximum variance (S) and, at the same time, they are intercorrelated with each other.

Thus, from the vector X, the covariance matrix, S, is obtained, given by

$$S = \frac{1}{n}\sum_{i=1}^{n}(x_i - \bar{x})^2. \tag{8.3}$$

Then, the eigenvectors $U = [u_1, u_2,..., u_p]$ associated with the eigenvalues of S with the highest magnitude are computed.

Having U, it is proceeded to obtain the new vector representation by the following expression:

$$y_1 = u_1^T X = u_{11}x_1 + \cdots + u_{k1}x_k + \cdots + u_{p1}x_p,$$

$$y_2 = u_2^T X = u_{12}x_1 + \cdots + u_{k2}x_k + \cdots + u_{p2}x_p,$$

$$\vdots$$

$$y_k = u_k^T X = u_{1k}x_1 + \cdots + u_{kk}x_k + \cdots + u_{pk}x_p, \tag{8.4}$$

$$\vdots$$

$$y_p = u_p^T X = u_{1p}x_1 + \cdots + u_{kp}x_k + \cdots + u_{pp}x_p,$$

subject to $u_1^T u_1 = 1, \ldots, u_p^T u_p = 1$. This restriction assures that the transformation is orthogonal, and the variance is not infinite. y_k corresponds to the k-dimensional reduced representation of the x vectors.

In this technique, variance plays a very important role, as it indicates the amount of information that incorporates each principal component. The higher the variance of the reduced variables, the more information of the original data retained by them (Jackson, 2003). Therefore, the component exhibiting the highest variance is selected as the first principal component, y_1. The second principal component is selected such that it is the one with the second highest variance but is not correlated to the first component. Similarly, the remaining components are obtained, ensuring that the pth component is orthogonal to all the eigenvectors previously calculated, i.e.,

$$u_p^T u_1 = u_p^T u_2 = \cdots = u_p^T u_{p-1} = 0. \tag{8.5}$$

In consequence, a limited number of the first components retain the greatest variance, so these components can replace the original variables, minimizing the size of the new set of variables, the principal components while preserving much of the original information. That is, the initial data group consisting of n observations and p variables is transformed to one consisting of n observations and k principal components, where $k \ll p$. That is the main benefit of the PCA: after a reduction in the dimensionality of the problem, a few variables retain the most important dynamics of the original data.

8.4 Curvilinear Component Analysis (CCA)

The CCA was proposed as an improvement to the Kohonen self-organizing maps (SOMs) (Demartines and Herault, 1997); it is a nonlinear mapping method, like multidimensional scaling and Sammon's nonlinear mapping. These methods minimize a cost function based on inter-point distances in both input and output spaces; nevertheless, the CCA algorithm applies a cost function to unfold closed data surfaces, a distinctive feature in nonlinear problems. The purpose is to expose a hidden pattern data using a low-dimension structure, as a pre-processing tool for clustering and classification processes.

The CCA will project a n-dimensional data set into a p-dimensional map (with $p \ll n$) with the aim to unfold nonlinear data structures and preserve the original distances of the data in the low-dimensional space. The CCA is implemented as a self-organizing one-layer neural network, each neuron unit has two weight vectors x_i and y_i; the number of elements in these vectors is the dimension of the input and the output space. In the SOM applications, vectors x_i are the prototypes, and the vectors y_i are fixed and transfer the position of the unit onto the grid. The principle of the CCA algorithm is shown in Figure 8.4. Learning starts with the vector quantization (VQ) on the input vectors and the random initialization of output vectors. The input weights first proceed to a VQ of the input data space, $\vec{X}$, in n dimensions. Then, the output weights map the local topology of the input average manifold by projecting it (P) into an output representation space, $\vec{Y}$, of dimension $p < n$. The output layer must build a nonlinear mapping of the input vectors, the Euclidean distances, X_{ij}, between x_i vectors are considered to do that. Corresponding distances in the output space are Y_{ij}.

To force Y_{ij} to match X_{ij} for each possible pair (i, j), the solution procedure minimizes iteratively the quadratic cost function:

$$E = \frac{1}{2}\sum_{i}\sum_{j \neq i}\left(X_{ij} - Y_{ij}\right)^2 F\left(Y_{ij}, \lambda_y\right), \tag{8.6}$$

where the weighting function, $F\left(Y_{ij}, \lambda_y\right)$, compensates the matching errors as the dimension is reduced from n to p. F is a bounded, monotonically decreasing function to local topology conservation and λ generally evolves with time (Demartines and Herault, 1997).

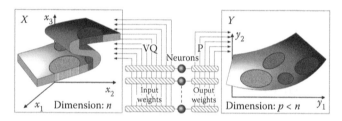

FIGURE 8.4
Principle of CCA.

8.5 PCA Applied to Discriminate Between Inrush and Fault Currents in Transformers

The PCA will extract the characteristic features from transformer differential current waveform during distinct scenarios; these features allow to classify the current signal as inrush transient (transformer energizing or over-excitation) or short circuit inside the power transformer.

The differential current in a power transformer is a three-phase signal; it is obtained from the current transformers placed on the primary and secondary sides of the protected transformer. Considering a Δ − Y, two-winding transformer, the differential error currents are given as follows:

$$I_{\text{DIFF}} = \begin{bmatrix} i_{\text{DIFF}(a-b)} \\ i_{\text{DIFF}(b-c)} \\ i_{\text{DIFF}(c-a)} \end{bmatrix} = \begin{bmatrix} I_{AB} - I_{ab} \\ I_{BC} - I_{bc} \\ I_{CA} - I_{ca} \end{bmatrix}. \tag{8.7}$$

where I_{AB}, I_{BC}, I_{CA}, and I_{ab}, I_{bc}, and I_{ca} are primary and secondary phase-to-phase currents, respectively. These currents pass through Delta filter (see Vazquez, et al., 2008) to eliminate any periodical components, such as load changes and harmonics. The output signal is called incremental differential current $\Delta i(t)$; they are computed by subtracting to the differential current, $i(t)$, the value corresponding to the preceding cycle, $i(t-T)$, as shown in Figure 8.5. With these signals, the sensitivity of the proposal differential algorithm is improved to detect any sudden changes of current due transient phenomena. Therefore, the algorithm does not require any setting to start its operation.

This filter subtracts, from any signal, the same signal with a time delay of one cycle. Thus, for a differential protection scheme, in the absence of any transient in the power system, the output of the filter is zero; otherwise, the filter output signal represents the occurring transient. This improves the sensitivity of the algorithm, reducing the effects of changes in the load that can undergo under normal system operation. Figure 8.6 shows the filter response for a main switch-opening event on the primary side of a power transformer occurring at a time t_0.

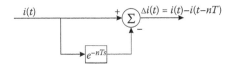

FIGURE 8.5
Delta filter structure used to obtain incremental signals from the differential currents.

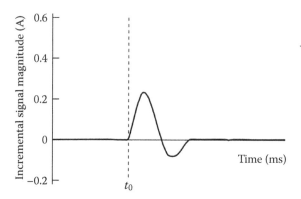

FIGURE 8.6
Delta filter response to a transformer's main switch opening event.

As mentioned above, the three-phase error differential current can be expressed as

$$\Delta I_{\text{DIFF}} = \begin{bmatrix} \Delta i_{\text{DIFF}(a-b)} \\ \Delta i_{\text{DIFF}(b-c)} \\ \Delta i_{\text{DIFF}(c-a)} \end{bmatrix}$$

$$= \begin{bmatrix} (I_{AB}(t) - I_{AB}(t-T)) - (I_{ab}(t) - I_{ab}(t-T)) \\ (I_{BC}(t) - I_{BC}(t-T)) - (I_{bc}(t) - I_{bc}(t-T)) \\ (I_{CA}(t) - I_{CA}(t-T)) - (I_{ca}(t) - I_{ca}(t-T)) \end{bmatrix}, \tag{8.8}$$

where T corresponds to a period of the signal at the system frequency, 16 ms for 60 Hz. During normal load condition, the incremental signals are zero. It reduces the effect of load changes in the power system.

In every instant of time, the input data to the algorithm is a half cycle of the three-phase incremental signal. Thus, the input matrix contains three incremental signals of differential current in half a cycle (64 samples), forming a 64×3-ΔI_T matrix:

$$\Delta I_T = \begin{bmatrix} \Delta i_{\text{DIFF}(a-b)1} & \Delta i_{\text{DIFF}(b-c)1} & \Delta i_{\text{DIFF}(c-a)1} \\ \vdots & \vdots & \vdots \\ \Delta i_{\text{DIFF}(a-b)64} & \Delta i_{\text{DIFF}(b-c)64} & \Delta i_{\text{DIFF}(c-a)64} \end{bmatrix}. \tag{8.9}$$

It is necessary to transpose ΔI_T since the purpose is to reduce the dimensions, so a 3×64 input matrix ΔI_T is obtained:

$$\Delta I_T = \begin{bmatrix} \Delta i_{\text{DIFF}(a-b)1} & \cdots & \Delta i_{\text{DIFF}(a-b)64} \\ \Delta i_{\text{DIFF}(b-c)1} & \cdots & \Delta i_{\text{DIFF}(b-c)64} \\ \Delta i_{\text{DIFF}(c-a)1} & \cdots & \Delta i_{\text{DIFF}(c-a)64} \end{bmatrix}. \tag{8.10}$$

The principal component transformation involves moving and rotating the data so that they are represented on a new plane. Hence, once the matrix ΔI_T is formed, the mean, $\overline{\Delta I_T}$, is subtracted from each data of the matrix, and it is divided by CT's secondary rated current (5 A), I_{TCr}, such that the matrix is normalized:

$$\Delta I_T^E = \frac{\Delta I_T - \overline{\Delta I_T}}{I_{\text{TCr}}}. \tag{8.11}$$

Having normalized the input matrix, it is proceeded to obtain the covariance matrix,

$$S = \begin{bmatrix} \text{cov}\{s_1, s_1\} & \text{cov}\{s_1, s_2\} & \cdots & \text{cov}\{s_1, s_{64}\} \\ \text{cov}\{s_2, s_1\} & \text{cov}\{s_2, s_2\} & \cdots & \text{cov}\{s_2, s_{64}\} \\ \vdots & \vdots & \ddots & \vdots \\ \text{cov}\{s_{64}, s_1\} & \text{cov}\{s_{64}, s_2\} & \cdots & \text{cov}\{s_{64}, s_{64}\} \end{bmatrix}, \tag{8.12}$$

whose size is 64×64. Then, the eigenvalues

$$L = \text{diag}\begin{bmatrix} \lambda_1 & \lambda_2 & \cdots & \lambda_{64} \end{bmatrix}, \tag{8.13}$$

and eigenvectors

$$U = \begin{bmatrix} \text{eig}_1 & \text{eig}_2 & \cdots & \text{eig}_{64} \end{bmatrix} \tag{8.14}$$

of S are computed.

The two dominant eigenvectors, corresponding to those associated with the two largest eigenvalues, are the new axes on which each column vector of ΔI_T will be transformed as follows:

$$\Delta I_{T(\text{PC})} = \begin{bmatrix} \text{eig}_1 & \text{eig}_2 \end{bmatrix}^T \Delta I_T^E, \tag{8.15}$$

where $\Delta I_{T(\text{PC})}$ is the transformed data matrix.

Thus, each of the differential currents, obtained by the presence of a disturbance in the system, is projected onto a two-dimensional subspace formed by the first two principal components. Based on this graphical representation, the discrimination between fault and inrush currents is carried out.

The differential protection based on PCA uses the qualitative performance of the differential current, and it is relatively insusceptible to variations in the electrical parameters of the transformer (maintenance maneuvers, aging). These changes modify the current signals during transformer connection, load transfers, or short circuits, but does not alter the physical phenomenon. Consequently, the waveforms will not be affected, especially if an appropriate method for scaling the signals is chosen.

8.6 Application of the CCA as a Base for a Differential Protection

Figure 8.7 shows the structure used for the neural network where DSP = digital signal processing stage, IL = input layer, OL = output layer, and DS = binary decision signal.

The IL input data consist of column vectors F_i; $i = 1,...,N$, with dimension 1×192. Those vectors are in turn formed by three vectors, $\Delta I_{\text{DIFF}(a-b)}$, $\Delta I_{\text{DIFF}(b-c)}$ and $\Delta I_{\text{DIFF}(c-a)}$, each with dimension 1×64, representing one cycle of the incremental differential current per phase. Each vector F_i represents

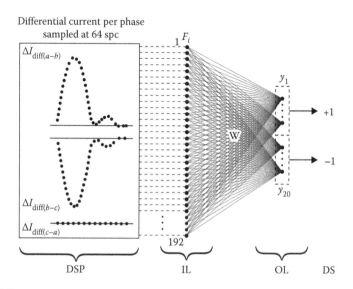

FIGURE 8.7
Structure used for the training of the artificial neural network.

a different power system operating scenario. The ANN is trained using these vectors.

The 192 elements stored on every single F_i vector are arranged in a training matrix T, with dimension $192 \times N$, where N represents the total number of test cases (340 cases in total):

$$T = \begin{bmatrix} F_{1,1} & F_{2,1} & \cdots & F_{340,1} \\ F_{1,2} & F_{2,2} & \cdots & F_{340,2} \\ \vdots & \vdots & \ddots & \vdots \\ F_{1,192} & F_{2,192} & \cdots & F_{340,192} \end{bmatrix}, \qquad (8.16)$$

with each $T_{i,j}$ element forming the training matrix satisfying the condition $0 \le T_{i,j} \le 1$.

Each column of T is presented to the network during the training stage on a random basis, so it does not affect how the data is organized.

The values of the connections between the 192 inputs and 20 output neurons are organized in a matrix of weights W, with dimension $192 \times N$:

$$W = \begin{bmatrix} W_{1,1} & W_{2,1} & \cdots & W_{340,1} \\ W_{1,2} & W_{2,2} & \cdots & W_{340,2} \\ \vdots & \vdots & \ddots & \vdots \\ W_{1,192} & W_{2,192} & \cdots & W_{340,192} \end{bmatrix}, \qquad (8.17)$$

with each $W_{i,j}$ element of W satisfying the condition $0 \le W_{i,j} \le 1$. At the beginning of the training, W is a null matrix.

The adaptation rule for the weights in W was based on the modified Kohonen adaptation rule (Kohonen, 1990). The choice of an output layer with 20 neurons was a heuristic decision based on the network training process.

The signals used as the algorithm's input are the three-phase differential current signals, i.e., $I_{\mathrm{DIFF}(a-b)}$, $I_{\mathrm{DIFF}(b-c)}$ and $I_{\mathrm{DIFF}(c-a)}$, and are stored in a matrix I_{DIFF}:

$$I_{\mathrm{DIFF}} = \begin{bmatrix} I_{\mathrm{DIFF}(a-b)} \\ I_{\mathrm{DIFF}(b-c)} \\ I_{\mathrm{DIFF}(c-a)} \end{bmatrix} = \begin{bmatrix} I_{AB} - I_{ab} \\ I_{BC} - I_{bc} \\ I_{CA} - I_{ca} \end{bmatrix}. \qquad (8.18)$$

Incremental current signals are obtained using a Delta filter (Vazquez, et al., 2008); it is the same filter uses in PCA approach and it is described in Figure 8.5.

Accordingly, the three-phase error differential current is computed as follows:

$$
\Delta I_{\text{DIFF}} = \begin{bmatrix} \Delta I_{\text{DIFF}(a-b)} \\ \Delta I_{\text{DIFF}(b-c)} \\ \Delta I_{\text{DIFF}(c-a)} \end{bmatrix}
$$

$$
= \begin{bmatrix} (I_{AB}(t) - I_{AB}(t-T)) - (I_{ab}(t) - I_{ab}(t-T)) \\ (I_{BC}(t) - I_{BC}(t-T)) - (I_{bc}(t) - I_{bc}(t-T)) \\ (I_{CA}(t) - I_{CA}(t-T)) - (I_{ca}(t) - I_{ca}(t-T)) \end{bmatrix}.
$$

(8.19)

A threshold value, ε, is defined, to serve as a criterion for starting the algorithm: $\Delta I_{\text{DIFF}} > \varepsilon$, where $\varepsilon = 5$ A, corresponding to the nominal secondary current of current transformers. When the incremental differential current exceeds the threshold value, ε, the algorithm begins to collect information necessary to form a cycle of the algorithm's input signals.

The incremental currents of the three phases are normalized by dividing each phase current value by the maximum phase value:

$$
\Delta I_{\text{DIFF}} = \begin{bmatrix} \dfrac{\Delta I_{\text{DIFF}(a-b)}}{\max\left\{\Delta I_{\text{DIFF}(a-b)}\right\}} \\[3ex] \dfrac{\Delta I_{\text{DIFF}(b-c)}}{\max\left\{\Delta I_{\text{DIFF}(b-c)}\right\}} \\[3ex] \dfrac{\Delta I_{\text{DIFF}(c-a)}}{\max\left\{\Delta I_{\text{DIFF}(c-a)}\right\}} \end{bmatrix}.
$$

(8.20)

According to the above equation, the algorithm input information corresponds to one cycle of the incremental signal simultaneously for each phase, sampled at 3.8 kHz (64 samples per cycle), i.e., a sampling time of 0.26 ms. Finally, it is obtained a column vector, $\Delta I_{\text{DIFF\_dis}}$, with dimension 192 × 1, representing the values of the three-phase incremental signal normalized, when it exceeds the threshold value ε:

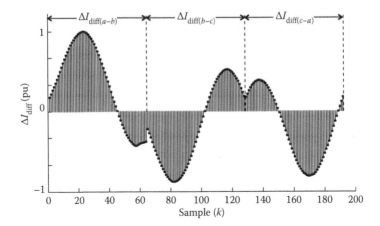

FIGURE 8.8
Incremental differential three-phase current signal for an internal three-phase short circuit at 50% of the transformer windings.

$$
\Delta I_{\text{DIFF\_dis}} =
\begin{bmatrix}
\Delta I_{\text{DIFF}(a-b)[1]} \\
\vdots \\
\Delta I_{\text{DIFF}(a-b)[64]} \\
\Delta I_{\text{DIFF}(b-c)[1]} \\
\vdots \\
\Delta I_{\text{DIFF}(b-c)[64]} \\
\Delta I_{\text{DIFF}(c-a)[1]} \\
\vdots \\
\Delta I_{\text{DIFF}(c-a)[64]}
\end{bmatrix}.
\tag{8.21}
$$

Figure 8.8 shows a 192-sample input signal; it is built with three cycles of incremental differential currents (one for each phase-to-phase current) for an internal three-phase fault at 50% of the windings, with a fault resistance of 1.0 Ω. In general, the algorithm control logic of the transformer differential protection using CCA is presented in Figure 8.9.

8.7 System Under Study

The test system consists of a three-phase, $\Delta - Y$, 100 MVA, 230/115 kV, and 60 Hz, transformer connected to a 230-kV line-to-line, three-phase

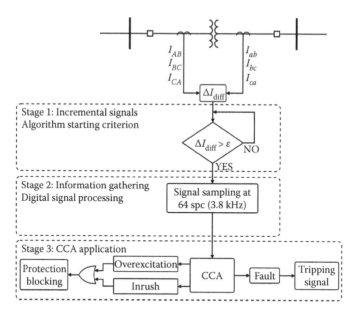

FIGURE 8.9
Proposed algorithm flowchart.

voltage source on its primary side with a 10 Ω source impedance, as shown in Figure 8.10.

The system includes a variable load module, a control module for faults in the transformer secondary, and a control module for short-circuit faults within the windings of the transformer. The source also has an *RL* impedance, which is modified to change the time constant of the dc aperiodic component. The excitation control module allows the transformer to work under sub- and over-excitation conditions. The system was simulated using the PSCAD®-EMTDC package.

The test system also has a frequency control to simulate the low-frequency conditions occurring, e.g., when the electric load connected to the system exceeds the capacity of generation.

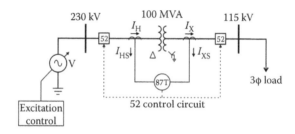

FIGURE 8.10
Test system diagram.

The current transformers used on both sides of the power transformer differ in their electrical behavior, i.e., have different magnetic saturation characteristics (Lucas, 1988). This results in an unbalanced operating current with an error proportional to both adjustment and transformation errors. During a transient condition, current transformers will behave differently. To avoid magnetic saturation in both cores, the following relation is considered (Altuve, et al., 2013):

$$\left(\frac{X}{R}+1\right)I_f Z_b \geq 20, \tag{8.22}$$

where X and R are the system reactance and resistance, respectively, seen from the transformer location, I_f is the maximum external fault current on the current transformer rated current base, and Z_b is the total burden on the current transformer rated burden base.

If the current transformers are selected according to (10.22), then none of them will reach the saturation region for the maximum fault outside the differential protection zone, thus reducing the differential current error. Tables 8.1 and 8.2 show the parameters of the current transformers used herein.

To obtain a wide range of incremental signals of the operating current, seven different scenarios were tested. Each of these scenarios was simulated

TABLE 8.1

Current Transformers' Parameter Data

Parameter	Units	Primary Side	Secondary Side
Magnetic hysteresis model		Joseph–Lucas	Jiles–Atherton
Primary turns		1	1
Secondary turns		200	60
Secondary resistance	Ω	0.61	0.5
Secondary inductance	mH	0.8	0.8
Area	m^2	6.501×10^{-3}	7.601×10^{-3}
Magnetic path length	m	0.5	0.6377
Remnant flux	T	0	0
Burden resistance	Ω	0.7	0.5
Burden inductance	Ω	0.9	0.8

TABLE 8.2

CTs' Transformation Ratio

Transformation Ratio	CT's Connection
300:5 (60)	Y (high-voltage side)
1000:5 (200)	Δ (low-voltage side)

at 17 different time instants within a cycle, for the effect of the disturbance time occurrence could be assessed. The considered scenarios are as follows: (1) no-load transformer energization, (2) on load transformer energization, (3) short circuit off the differential protection zone with no-load transformer energization, (4) short circuit off the differential protection zone with on load transformer energization, (5) short circuit within the differential protection zone with no-load transformer energization, (6) short circuit within the differential protection zone with on load transformer energization, and (7) a mix of the above scenarios.

8.8 Results

8.8.1 Discussion of the Principal Component Analysis Results

Applying the PCA technique permits the input matrix composed of normalized differential currents, to be transformed from 64 dimensions into two dimensions. The thresholds to discriminate inrush currents from fault currents were defined heuristically for the new data coordinates. Table 8.3 summarizes the obtained thresholds, which were consistent in all simulations.

On the new two-dimensional space defined by the first two principal components, if the projection of the three-phase currents over the first principal component PC1 is within the threshold [−0.04, 0.04], it is determined that the occurred event is an energization or over-excitation condition (inrush current). Otherwise, if the projection is outside the above threshold, i.e., $(-\infty, -0.04][0.04, \infty)$, it is determined that the currents are due to the presence of an internal fault and the differential relay operation should be allowed.

Figure 8.11 shows the incremental differential currents obtained from the initial connection of the transformer and its projection in PC. It is seen that the three projections are within the specified threshold for an inrush condition.

The projections of the differential currents, for the case of 110% over-excitation, are shown in Figure 8.12(b). Figure 8.13(b) shows the projection into PC of incremental differential currents recorded from a single-phase fault present inside the protection zone of the transformer. It is observed that the differential currents affected by this failure exceed the specified inrush

TABLE 8.3

Operating Thresholds for a Two-Winding Transformer PCA-based Differential Protection

Event	PC1
Energization or over-excitation	[−0.04, 0.04]
Internal fault	$(-\infty, -0.04][0.04, \infty)$

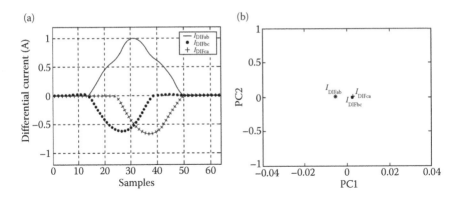

FIGURE 8.11
Differential currents due to an energization and its projection into the PC1–PC2 subspace.

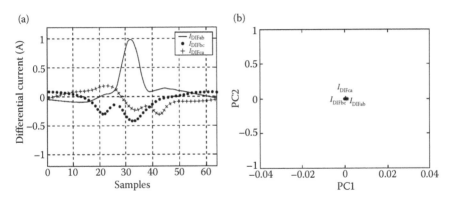

FIGURE 8.12
Differential currents due to a 110% over-excitation and its projection into the PC1–PC2 subspace.

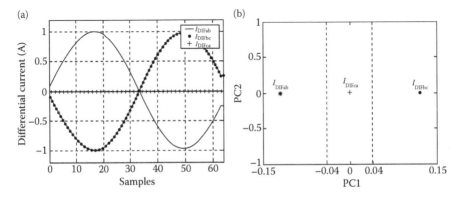

FIGURE 8.13
Differential currents due to a single-phase fault during transformer energization and its projection into the PC1–PC2 subspace.

condition threshold, while the current that remains unaffected by the failure is projected within the threshold.

For the case of an internal fault in the transformer (short circuit between the phase C windings), the projection of the differential currents into PC is shown in Figure 8.14(b). Since at least one of the three projected values is out of the [−0.04, 0.04] range, it is confirmed that this is an inrush condition.

Nonlinear load introduces distortion in current signals; this scenario will be considered in the test system for evaluating the performance of the proposed algorithm. A six-pulse power electronic converter is connected as a load on the secondary side of the protected transformer, it produces a total harmonic distortion (THD) level of 28.45%, approximately. Despite this distortion level is too high, no filter or device is used to reduce the distortion levels below those permitted by the standards to verify the algorithm performance in this critical scenario. Figures 8.15 and 8.16 show the projection

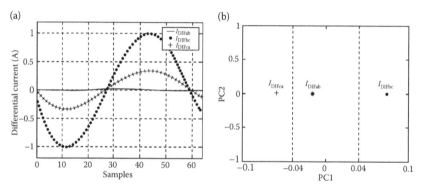

FIGURE 8.14
Differential currents due to a transformer's internal fault and its projection into the PC1–PC2 subspace.

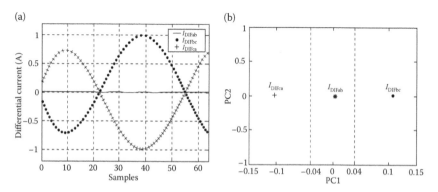

FIGURE 8.15
Differential currents due to a single-phase fault and its projection into the PC1–PC2 subspace.

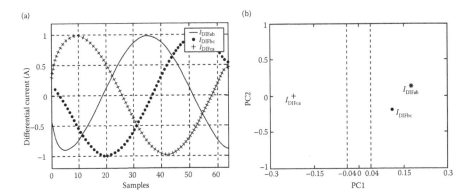

FIGURE 8.16
Differential currents due to a three-phase fault and its projection into the PC1–PC2 subspace.

into PC for this scenario in the case of a fault occurs. Again, it is noted that the projection corresponding to currents, which are affected by the fault, exceeds the [−0.04, 0.04] threshold.

The magnitude of inrush current of the transformer depends on several factors, such as transformer capacity and its saturation curve. Then, a new simulation was carried out in the same test system using a 25-MVA transformer and a different saturation curve from the hitherto managed, yielding the projection shown in Figure 8.17.

Additionally, the connection used for the transformer has been a Δ − Y. To demonstrate that the proposed algorithm is immune to the transformer connection type, its performance was evaluated for the protection of a Y − Δ transformer, obtaining the projection shown in Figure 8.18.

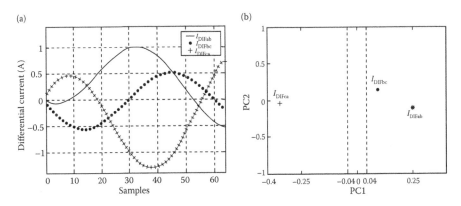

FIGURE 8.17
Differential currents due to a phase-to-phase fault and its projection into the PC1–PC2 subspace.

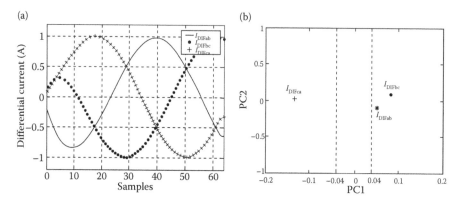

FIGURE 8.18
Differential currents due to a three-phase fault and its projection into the PC1–PC2 subspace.

8.8.2 Discussion of the Curvilinear Component Analysis Results

The complete training matrix consists of 192 rows (input information to the network) and 340 columns or simulated cases (65,280 data points). Of these, 170 cases fall within the cases where the differential protection should not be blocked (fault occurs within the protection zone), while the remaining 170 cases are those where differential protection should be blocked (fault occurs outside the protection zone). Table 8.4 details the training matrix structure. The total time for the network to be trained was 15 s, with a total of 100 iterations. Figure 8.19 shows the output of the network once it has been trained. It can be seen the correct discrimination between faults outside the protection zone (+1 for blocking vectors in Table 8.4) and faults within the protection zone (−1 for operation vectors in Table 8.4).

To test the performance of the network, new test vectors, not used before to train the network, were employed. To this end, four new scenarios were simulated to test the effectiveness of the algorithm. For each scenario, 17 different instants of time were also simulated. Thus, obtaining a total of 68 cases that were stored in a test matrix, T, with dimension 192×68.

The algorithm was tested under the following conditions: energizing a power transformer with another one already in service (see Figure 8.20); overexciting the original transformer, to a 135% of its rated primary voltage (see Figure 8.21); a single phase to ground fault. For the last case, the fault occurred at 10% and 90% of the phase winding (see Figure 8.22).

TABLE 8.4

Vectors Forming the Test Matrix

Blocking Vectors	Operating Vectors
1–34	35–68

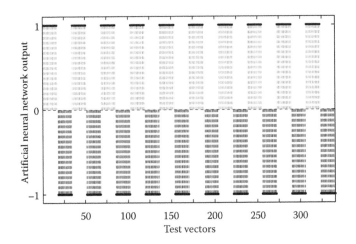

FIGURE 8.19
Artificial neural network training output.

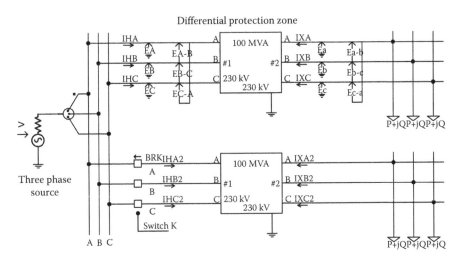

FIGURE 8.20
Connecting a transformer in parallel with another already in service using the switch K.

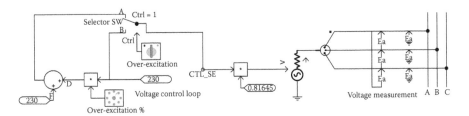

FIGURE 8.21
Transformer's primary voltage control loop.

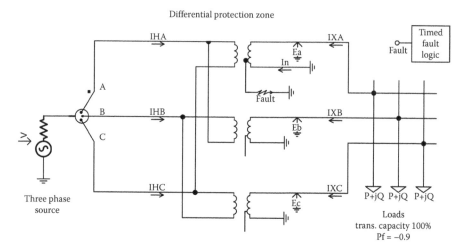

FIGURE 8.22
Control scheme for internal faults.

Figure 8.23 shows the results obtained when the proposed algorithm is applied to a completely new test matrix, which has never previously been presented to the network. The ANN achieves proper discrimination between transformer inrush currents and short circuits within the protection zone currents; the first group of 34 outputs are the blocking scenarios and the second group outputs are the operating scenarios.

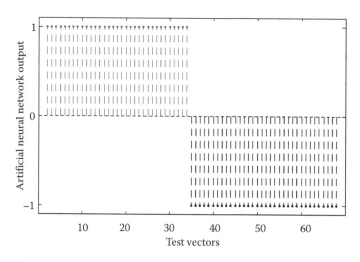

FIGURE 8.23
Artificial neural network response to the test matrix.

8.9 Conclusion

The proposed differential protection algorithms are based on the principal component and curvilinear component analysis techniques for the discrimination between inrush and short-circuit transformer currents. This approach uses the projection of the differential error currents into reduced dimensional spaces, so the process of identifying the event occurred to the system is simplified.

Although to validate the algorithm performance, many simulations were carried out, for different types of event, only a few of them were shown in the chapter. Therefore, in every test, certain parameters and characteristics of the transformer and system that directly affect the inrush current were modified, analyzing the algorithms' behavior when subjected to different scenarios.

The results were satisfactory for the PCA and CCA algorithms; they achieved a correct discrimination of the various events simulated in the test power system. The results described in this chapter proved that both algorithms have high reliability and security; the main advantage of them is their simplicity and fast operation time.

References

H. Abniki, H. Afsharirad, A. Mohseni, F. Khoshkhati, H. Monsef, and P. Sahmsi, Adaptive harmonic estimation technique for reduction the blocking time of transformer for differential protection, in *Proceedings of the 2010 North American Power Symposium*, Arlington, p. 208.

H.J. Altuve, N. Fischer, G. Benmouyal, and D. Finney, Sizing current transformers for line protection applications, in *Proceedings of the 66th Annual Conference for Protective Relay Engineers*, College Station, Apr. 2013.

A. Ashrafian, M. Mirsalim, and M.A.S. Masoum, Application of a recursive phasor estimation method for adaptive fault component based differential protection of power transformers, *IEEE Transactions on Industrial Informatics*, vol. 13, no. 3, pp. 1381–1392, 2017.

K. Bunte and J.A. Lee, Unsupervised dimensionality reduction: The challenges of big data visualization, in *Proceedings of the European Symposium on Artificial Neural Networks, Computational Intelligence and Machine Learning*, pp. 487–494, Bruges, Belgium, 2015.

R.E. Cordray, Percentage-differential transformer protection, *Electrical Engineering*, vol. 50, no. 5, pp. 361–363, 1931.

P. Demartines and J. Herault, Curvilinear component analysis: a self-organizing neural network for nonlinear mapping of data sets, *IEEE Transactions on Neural Networks*, vol. 8, no. 1, pp. 148–154, 1997.

W.R. Dillon and M. Goldstein, *Multivariate Analysis: Methods and Applications*, New York: John Wiley, 1984.

A. Guzman, S. Zocholl, G. Benmouyal, and H.J. Altuve, A current-based solution for transformer differential protection Part II: relay description and evaluation, *IEEE Power Engineering Review*, vol. 22, no. 7, pp. 60–60, 2002.

T. Heskes, Energy functions for self-organizing maps, in E. Oja and S. Kaski, ed., *Kohonen Maps*, pp. 303–315, Amsterdam: Elsevier, 1999.

H. Hotelling, Analysis of a complex of statistical variables into principal components, *Journal of Educational Psychology*, vol. 24, no. 6, pp. 417–441, 1933.

J.E. Jackson, *A User's Guide to Principal Components*, Hoboken, NJ: Wiley, 2003.

I.T. Jolliffe, *Principal Component Analysis*, 2nd ed. New York: Springer-Verlag, 2002.

M. Kezunovic, L. Xie, and S. Grijalva, The role of big data in improving power system operation and protection, in *Proceedings of the IX IREP Symposium -Bulk Power Systems Dynamics and Control*, pp. 1–9, Rethymnon, Greece, 2013.

E. Kiliç, O. Özgöönenel, O. Usta, and D. Thomas, PCA based protection algorithm for transformer internal faults, *Turkish Journal of Electrical Engineering & Computer Sciences*, vol. 17, no. 2, pp. 125–142, 2009.

T. Kohonen, The self-organizing map, *Proceedings of the IEEE*, vol. 78, no. 9, pp. 1464–1480, 1990.

J.A. Lee and M. Verleysen, *Nonlinear Dimensionality Reduction*, New York: Springer, 2007.

J.R. Lucas, Representation of magnetization curves over a wide region using a non-integer power series, *International Journal of Electrical Engineering Education*, vol. 25, no. 4, pp. 335–340, 1988.

O.P. Malik, P.K. Dash, and G.S. Hope, Digital protection of a power transformer, in *Proceedings of the 1976 IEEE PES Winter Meeting*, New York, 1976.

K. Pearson, On lines and planes of closest fit to systems of points in space, *Philosophical Magazine*, series 6, vol. 2, no. 11, pp. 559–572, 1901.

A.G. Phadke and J.S. Thorp, *Computer Relaying for Power Systems*, 2nd ed. Chichester: John Wiley & Sons, 2009.

M.A. Rahman, Advancements in digital protection of power transformers, in *Proceedings of the 2006 IEEE International Power and Energy Conference*, Putra Jaya, Malaysia, Nov. 2006.

M. Tripathy, Power transformer differential protection based on neural network principal component analysis, harmonic restraint and Park's plots, *Advances in Artificial Intelligence*, vol. 2012, pp. 1–9, 2012.

E. Vazquez, I.I. Mijares, O.L. Chacon, and A. Conde, Transformer differential protection using principal component analysis, *IEEE Transactions on Power Delivery*, vol. 23, no. 1, pp. 67–72, 2008.

O. Youssef, A wavelet-based technique for discrimination between faults and magnetizing inrush currents in transformers, *IEEE Transactions on Power Delivery*, vol. 18, no. 1, pp. 170–176, 2003.

T. Zhang and B. Yang, Big data dimension reduction using PCA, in *Proceedings of the 2016 IEEE International Conference on Smart Cloud*, pp. 152–157, New York, 2016.

Index

A

ADC devices. *see* Analog-to-digital (ADC) converters
ADMM. *see* Alternating direction method of multipliers (ADMM)
Advanced communication network, 40
Advanced metering infrastructure (AMI), 39, 43, 44, 46, 48–50
Algorithms, for big data
 deep learning, 19–20
 for high variety of data, 26
 for high-velocity of data, 26
 for high-volume of data, 25–26
 models, 20–25
 machine learning, 13–14
 artificial neural network, 14–15
 decision-tree classifier, 17–19
 support vector machine, 15–17
Alternating direction method of multipliers (ADMM), 69
AMI. *see* Advanced metering infrastructure (AMI)
AMR. *see* Automatic meter reading (AMR)
Analog-to-digital (ADC) converters, 113
ANN. *see* Artificial neural network (ANN)
Artificial neural network (ANN), 14–15, 22, 29
Automatic meter reading (AMR), 114

B

Bandwidth, 40, 44
BD. *see* Benders Decomposition (BD)
Benders Decomposition (BD), 32
Big data
 algorithms for processing (*see* Algorithms for big data)
 applications, 10–11
 hardware capability, 13
 health care, 11–12
 in power systems, 27–33
 social networking, 12–13
 decomposition methodology handling, 31–32
 definitions, 2, 56
 in large-scale power system, 27–33
 optimization and techniques, 64–71
 overview, 10
 and power system protection, 143–146 (*see also* Power transformer differential protection)
 scientometric analysis of, 58–63
 technology, 41
Big data set, 41, 46, 51
Biometric security, 98–99
Boudle method, 67
Bundle method, 67

C

CART algorithm, 17–18
CCA. *see* Curvilinear component analysis (CCA)
CI. *see* Critical infrastructure (CI)
CIM. *see* Common information model (CIM)
CNN. *see* Convolutional neural network (CNN)
Commercial Off the Shelf (COTS), 91
Commercial technology, 91–92
Common information model (CIM), 29–30
Communication infrastructure, 40, 41, 48
Communication security, 87
Communication system threats
 characteristics of, 92
 commercial technology inroads into, 91–92
 data acquisition in, 89
 data damage-related threats, 95
 device property issues, 93–94
 effects of, 87

Communication system threats (*cont.*)
    existential-related issues, 93–94
    host-based threats, 94
    insider *versus* outsider threats, 92–93
    operations and components, 89–91
    physical *versus* electronic threats,
        94–95
    stack-based exploitations, 95–96
    supply-chain-related threats, 95
Conjugate gradient methods, 66
Convex optimization, 68–69
Convolutional neural network (CNN),
    20–22
COTS. *see* Commercial Off the Shelf
    (COTS)
Critical infrastructure (CI)
    communication security, 87
    communication system threats and
        operations, 87–89
    cyber threats, 96–101
    high-level communication system
        threats, 92–96
    industrial control networks and
        operations, 89–92
    overview, 86–87
    security, 96–101
Curvilinear component analysis (CCA),
    145, 149–150, 154–157, 164–166,
    167
Cyber threats and security, 40
    component-specific-related threats,
        97
    physical-layer threats and security
        measures, 98–101
    software and communication threats,
        97–98

D

DADP. *see* Dual Approximate Dynamic
    Programming algorithm
Dantzig–Wolfe decomposition method,
    31
Data access, 57
Data analysis, 57
Data collaboration, 57
Data damage threats, 95, 96
Data mining, electricity theft, 116–117
    classification and clustering, 117–118

detection, 118
    issues and directions in, 118–120
    prediction, 117
Data preparation, 57
Data reporting, 57
Data throughput, 40
Data traffic pattern
    smart grid environment in, 42–43
        advanced metering infrastructure,
            44
        phasor measurement unit, 43–44
Data validation, 57
DBN. *see* Deep belief network (DBN)
DCSs. *see* Distributed Control Systems
Decision-making method, 136–138
Decision-tree classifier, 17–19
Deep belief network (DBN), 24–25
Deep learning
    for high variety of data, 26
    for high-velocity of data, 26
    for high-volume of data, 25–26
    models, 20–25
Demand response (DR), 43, 45
Digital data, 12
Digital meters, electricity theft, 114, 116
Digital signal processor (DSP), 113–114
Distributed Control Systems (DCSs), 89
Distributed energy resources, 45
Distribution automation, 46
DOE. *see* U.S. Department of Energy
    (DOE)
DR. *see* Demand response (DR)
DSP. *see* Digital signal processor (DSP)
Dual Approximate Dynamic
    Programming algorithm
    (DADP), 32

E

Economic dispatch (ED), 69
Electrical power systems, 4
    application of big data, 58, 65
    big data optimization and
        techniques, 64–71
    overview, 56
    scientometric analysis, 58–63
Electrical vehicle (EV), 134–135
Electricity markets. *see* Financial
    transmission rights (FTRs)

Electricity theft, 4, 120
   billing issues, 114–115
   and data collection, 116
   data mining and (*see* data mining, electricity theft)
   fraud, 111–114
   outright theft, 115–116
   overview, 107–108
   transmission and distribution system losses, 108–110
Electric Power Research Institute (EPRI), 30
Electric power system, 38
Energy production/consumption forecast, 139
ENTSO-E. *see* European Network of Transmission System Operators (ENTSO-E)
EPRI. *see* Electric Power Research Institute (EPRI)
European Network of Transmission System Operators (ENTSO-E), 29
EV. *see* Electrical vehicle (EV)

F

FAHCLPSO, 71
Financial transmission rights (FTRs), 32–33
Fossil fuels, 126
4Vs data, 10. *see also* Big Data
FTRs. *see* Financial transmission rights (FTRs)

G

Gini gain, 17–19
Gini Index, 18
Global positioning system (GPS), 28–29
Grid modernization, 38–39
Grid monitoring, 42

H

Hardware capability, 13
Harmonic blocking methods, 146–147
Harmonic restraint methods, 146–147
Health care analytics, 11–12

Hestenes–Stiefel (HS) method, 66
HMI. *see* Human machine interfaces (HMI)
Host-based threats, 94
HS method. *see* Hestenes–Stiefel (HS) method
Human machine interfaces (HMI), 89
Hybrid algorithm, 70

I

ICTs. *see* Information and Communication Technologies (ICTs)
IDPSs. *see* Incorporating intrusion detection and prevention systems (IDPSs)
IEC. *see* International Electrotechnical Commission (IEC)
IEDs. *see* Intelligent electronic devices (IEDs)
IEEE 123 distribution systems, 47
Incorporating intrusion detection and prevention systems (IDPSs), 97
Industrial control networks, 91–92
Information and Communication Technologies (ICTs), 27, 38
Insider threats, 93
Intelligent communication network, 39, 40
Intelligent electronic devices (IEDs), 38
International Electrotechnical Commission (IEC), 30
Internet, 10–11
Internet of Things (IoT), 40
Interoperability, 40
IoT. *see* Internet of Things (IoT)

J

Java Development (JaDe), 133

L

Large-scale power system
   big data problem, 30–33
   citation analysis, 61
   common information model standard, 29–30
   phasor measurement units, 28–29

Large-scale power system (*cont.*)
  renewable energy, 29
  smart grid networks, 27–28
Large-scale unconstrained
    optimization, 66
Latency, 40, 44
Linear programming (LP), 31, 32
Line outage distribution factors
    (LODFs), 30
Logistics
  network, 67
  optimization, 67–68
Long short-term memory (LSTM), 23
LP. *see* Linear programming (LP)
LSTM. *see* Long short-term memory
    (LSTM)

M

Machine learning, 4
  artificial neural network model, 14–15
  decision-tree classifier, 17–19
  support vector machine, 15–17
  types of, 13–14
Metaheuristic algorithms, 69–71
Meter tampering, 111–114
MLP. *see* Multi-layer perceptron (MLP)
Mobile technologies, 42
Multi-agent architecture, 133–134.
        see also Smart grids (SG)
Multi-layer perceptron (MLP), 14

N

Narrow band IoT (NB-IoT) networks,
    42, 48
NASPI. *see* North American
    SynchroPhasor Initiative
    (NASPI)
NB-IoT networks. *see* Narrow band IoT
    (NB-IoT) networks
Network-based routing attacks, cyber
    threats, 97
Nonconvex optimization, 68–69
Nonsmooth large-scale optimization, 67
Nontechnical losses (NTLs), 108–110
North American SynchroPhasor
    Initiative (NASPI), 29
NTLs. *see* Nontechnical losses (NTLs)

O

Open Systems Interconnection (OSI)
    model, 88, 96
Orthogonal vectors, 148, 149
OSI model. *see* Open Systems
    Interconnection (OSI) model
Outright electricity theft, 115–116
Outsider threats, 92–93, 94

P

Particle swarm optimization (PSO),
    69, 71
PCA. *see* Principal component analysis
    (PCA)
PH. *see* Progressive Hedging (PH)
Phasor measurement units (PMU), 27,
    28–29, 39, 43–44, 48
Photovoltaic (PV) cells. *see* solar cells
Physical-layer and security measures
  biometric-like security with, 98–99
  communication signal exploitation,
    100–101
  physically traceable objects, 99–100
Physical Unclonable Functions (PUFs),
    99
PLCs. *see* Programmable Logic
    Controllers (PLCs)
PMU. *see* Phasor measurement units
    (PMU)
P–N junction, 128
Pooling layers, 21, 22
Power system modeling, big data
    problem in, 30–33
Power transformer differential
    protection, 145
  curvilinear component analysis,
    149–150, 154–157, 164–166
  harmonic restraint and blocking
    methods, 146–147
  principal component analysis,
    148–149, 151–154, 167
  test system analysis, 157–160
  waveform recognition, 147–148
Principal component analysis (PCA),
    145, 148–149, 151–154
Programmable Logic Controllers
    (PLCs), 90, 97

Progressive Hedging (PH), 32
PSO. *see* Particle swarm optimization
    (PSO)
PUFs. *see* Physical Unclonable Functions
    (PUFs)

R

Radio Frequency Identification (RFID), 99
Raw data, 12
RBMs. *see* Restricted Boltzmann
    machines (RBMs)
Recurrent neural networks (RNNs), 23
Remote Telemetry Units (RTUs), 90
Renewable power energy, 29, 126
    solar power, 128–131
    wind power, 126–128
Restricted Boltzmann machines (RBMs),
    24, 25
RF Certificates of Authenticity
    (RF-COA), 99
RF fingerprinting, 100–101
RFID. *see* Radio Frequency Identification
    (RFID)
RNNs. *see* Recurrent neural networks
    (RNNs)
Robust communication infrastructure,
    40, 41
RTUs. *see* Remote Telemetry Units
    (RTUs)

S

SCADA. *see* Supervisory Control And
    Data Acquisition (SCADA)
Scientometric analysis, 58–63
Security-constrained Unit Commitment
    (SCUC), 30–31
Self-organizing maps (SOMs), 149
Smart grids (SG)
    agent's profile, 135–136
    data traffic pattern in, 42–44
    decision-making method, 136–138
    description, 4, 37–38
    extra power procedure, 138–139
    interconnection with internet of
        things, 39–42
    massive flow of information in, 45–46
    modernization, 38–39

multi-agent system, 134–135
networks, 27–28
volume of generated data (case of
    study), 47–50
Social networking, 12–13
Software threats, 97–98
Solar cells, 128
Solar panels, 129
    capacity, 129–130
    efficiency, 130
    power collected *versus* energy
        collected, 130–131
    power generation density, 130
Solar power, 128–131
SOMs. *see* Self-organizing maps (SOMs)
Stack-based exploitations, 95–96
Subsampling layers, 21
Supervised learning, 13–14
Supervisory Control And Data
    Acquisition (SCADA), 28, 89
Supply-chain threats, 95
Support vector machine (SVM)
    kernel function, 17
    regression problem, 15–16
SVM. *see* Support vector machine (SVM)
Synchrophasor technology, 28

T

TC 57. *see* Technical Committee 57 (TC 57)
T&D system losses. *see* Transmission
    and Distribution (T&D) system
    losses
Technical Committee 57 (TC 57), 30
Technical losses (TL), 108–110
Theft. *see* Electricity theft
Time-synchronized phasor. *see* Phasor
    measurement units (PMU)
TL. *see* Technical losses (TL)
Transmission and Distribution (T&D)
    system losses, 108–110
Two-way communication system, 39

U

UC Consumer Plan (UCCP), 136–138
UCPP. *see* UC Producer Plan (UCPP)
UC problem. *see* Unit commitment (UC)
    problem

UC Producer Plan (UCPP), 136–138
Unconstrained optimization, 66
Unit commitment (UC) problem, 4,
      125–126, 131–133, 140
Unsupervised learning, 14
U.S. Department of Energy (DOE), 29
US health care system, 12

V

Vast sensor network, 37
Vast smart distribution network, 43

W

WAMS. *see* Wide-area
      measurement systems
      (WAMS)
Waveform recognition, 147–148
Weighing parameters, 13
Wide-area measurement systems
      (WAMS), 27
Wide area monitoring system, 42
Wind power, 126–127
Wind turbine, 127–128